ASE Test Preparation Series

Transit Bus Technician Test

Diesel Engines

(Test H2)

1st Edition

DELMAR
CENGAGE Learning

Australia • Brazil • Japan • Korea • Mexico • Singapore • Spain • United Kingdom • United States

DELMAR
CENGAGE Learning

ASE Test Preparation Series: Transit Bus Technician Test: Diesel Engines (Test H2), First Edition

Vice President, Technology Professional Business Unit: Gregory L. Clayton

Director, Professional Automotive and Trucking Learning Solutions: Kristen L. Davis

Product Manager: Kimberley Blakey

Editorial Assistant: Jason Yager

Director of Marketing: Beth A. Lutz

Marketing Coordinator: Jennifer Stall

Production Director: Patty Stephan

Production Manager: Andrew Crouth

Content Project Manager: Kara A. DiCaterino

Art Director: Robert Plante

Cover Design: Michael Egan

ISBN-13: 978-1-4180-6570-6

ISBN-10: 1-4180-6570-6

Delmar
Executive Woods
5 Maxwell Drive
Clifton Park, NY 12065
USA

Cengage Learning is a leading provider of customized learning solutions with office locations around the globe, including Singapore, the United Kingdom, Australia, Mexico, Brazil, and Japan. Locate your local office at **www.cengage.com/global**

Cengage Learning products are represented in Canada by Nelson Education, Ltd.

To learn more about Delmar, visit **www.cengage.com/delmar**

Purchase any of our products at your local bookstore or at our preferred online store **www.ichapters.com**

Printed in the United States of America
3 4 5 6 7   15 14 13 12 11

ED083

Contents

Section 5 Sample Test for Practice

Section 6 Additional Test Questions for Practice

Section 7 Appendices

Preface

Delmar, Cengage Learning is very pleased that you have chosen our ASE Test Preparation Series to prepare yourself for the Transit Bus Diesel Engines (H2) ASE Examination. This guide is designed to introduce you to the Task List for the Diesel Engines (H2) test you are preparing to take, give you an understanding of what you are expected to be able to do in each task, and take you through sample test questions formatted in the same way the ASE tests are structured.

If you have a basic working knowledge of the discipline you are testing for, you will find Delmar's ASE Test Preparation Guide to be an excellent way to understand the "must know" items to pass the test. This book is not a textbook. Its objective is to prepare the technician who has the requisite experience and schooling to challenge ASE testing. It cannot replace the hands-on experience or the theoretical knowledge required by ASE to master vehicle repair technology. If you are unable to understand more than a few of the questions and their explanations in this book, it could be that you require either more shop-floor experience or further study.

The book begins with an item-by-item overview of the ASE Task List with explanations of the minimum knowledge you must possess to answer questions related to the task. Following the overview are two sets of sample questions, answer keys, and explanations of the answers to each question. A few of the questions are not strictly ASE format but are included because they help teach a critical concept that will appear on the test. We suggest that you read the complete Task List Overview before taking the first sample test. After taking the first test, score yourself and read the explanation to any questions that you were not sure about, including the questions you answered correctly. Each test question has a reference to the related task or tasks that it covers so you can go back and read over any area of the Task List that you are having trouble with. Once you are satisfied that you have all of your questions answered from the first sample test, take the additional test and check it. If you pass these tests, you will do well on the ASE test.

Our Commitment to Excellence

Delmar, Cengage Learning has sought out the best technicians in the country to help with the development of this 1st edition of the Transit Bus Diesel Engines (H2) ASE Test Preparation Guide.

Thank you for choosing Delmar's ASE Test Preparation Series. All of the writers, editors, and Delmar staff have worked very hard to make this Test Preparation Guide second to none. We know you are going to find this book accurate and easy to work with. It is our ongoing objective to improve our products at Delmar by responding to feedback. If you have any questions concerning the books in this series, email us at: autoexpert@trainingbay.com.

1 The History and Purpose of ASE

ASE began as the National Institute for Automotive Service Excellence (NIASE). It was founded as a nonprofit independent entity in 1972 by a group of industry leaders with the single goal of providing a means for consumers to distinguish between incompetent and competent mechanics. ASE accomplishes this goal by testing and certifying repair and service professionals. From this beginning it has evolved to be known simply as ASE (Automotive Service Excellence) and currently offers more than 40 certification exams in automotive, medium/heavy-duty truck, collision, engine machinist, school bus, parts specialist, automobile service consultants, and other industry-related areas.

Today, more than 400,000 professionals hold current ASE certifications. These technicians are employed by new car and truck dealerships, independent garages, fleets, service stations, franchised service facilities, and more. ASE continues its mission by also providing information to help consumers identify repair facilities that employ certified professionals through its Blue Seal of Excellence Recognition Program. Shops that employ a minimum of 75 percent ASE-certified repair technicians and meet other criteria can apply for and receive the Blue Seal of Excellence Recognition.

ASE recognized that educational programs serving the service and repair industry also needed a way to be recognized as having the faculty, facilities, and equipment to provide a quality education to students wanting to become service professionals. Though the combined efforts of the ASE and industry and education leaders, the nonprofit National Automotive Technicians Education Foundation (NATEF) was created to evaluate and recognize training programs. Today more than 2000 programs are ASE certified under the standards set by the service industry. In addition, ASE/NATEF offers certification of industry (factory) training programs through Continuing Automotive Service Education (CASE). CASE recognizes training provided by replacement parts manufacturers as well as vehicle manufacturers.

ASE certification testing is administered by the American College Testing (ACT) service. Strict standards of security and supervision at the test centers ensure that the technician who holds the certification earned it. In addition, ASE certification requires that technicians passing the test demonstrate that they have two years of work experience in the field before they can be certified. Industry experts who are actually working in the field being tested develop test questions. More detailed information on how the test is developed and administered is provided in the next section. Paper-and-pencil tests are administered twice a year at more than 750 locations in the United States. Computer Based Testing (CBT) is now also available with the benefit of instant test results at certain established test centers. ASE certification is valid for 5 years and can be recertified by retesting.

So that consumers can recognize certified technicians, ASE issues a jacket patch, certificate, and wallet card to certified technicians and makes signs available to facilities that employ ASE-certified technicians. You can contact ASE at:

National Institute for Automotive Service Excellence
101 Blue Seal Drive S.E.
Suite 101
Leesburg, VA 20175
Telephone 703-669-6600
FAX 703-669-6123
www.ase.com

WE SUPPORT
PROFESSIONAL CERTIFICATION
THROUGH THE
National Institute for
AUTOMOTIVE
SERVICE
EXCELLENCE

2 Take and Pass Every ASE Test

Participating in an Automotive Service Excellence (ASE) voluntary certification program gives you a chance to show your customers that you have the "know-how" needed to work on today's modern vehicles. ASE certification tests allow you to compare your skills and knowledge to the automotive service industry's standards for each specialty area.

If you are the "average" automotive technician taking this test, you are in your mid-30s and have not attended school for about 15 years. That means you probably have not taken a test in many years. Some technicians, on the other hand, have attended college or taken postsecondary education courses and may be more familiar with taking tests and with test-taking strategies. There is, however, a difference in the ASE test you are preparing to take and the educational tests you may be accustomed to.

How Are the Tests administered?

ASE tests are administered at more than 750 test sites in local communities. Paper-and-pencil tests are the type most widely available to technicians. Each tester is given a booklet containing questions with charts and diagrams where required. You can mark in this test booklet but no information entered in it is scored. Answers are recorded a separate answer sheet. You enter your answers, in number 2 pencil only. ASE recommends you bring 4 sharpened number 2 pencils with erasers. Answer choices are recorded by marking (bubble in) a block by the question number on the answer sheet. These sheets are scanned electronically and the answers tabulated. For test security booklets include randomly generated questions. Your answer key must be matched to the proper booklet so it is important to enter the booklet serial number correctly on the answer sheet. All instructions are printed on the test materials and should be followed carefully.

ASE has introduced Computer Based Testing (CBT) at some locations. While test content is the same for both testing methods, CBT tests have some unique requirements and advantages. If you are considering the CBT tests, visit the ASE web page (www.ASE.com), and review the conditions and requirements for this type of test. A test demonstrator page allows you to experience this type of test before you register. Some technicians find that this style of testing provides an advantage while others find operating the computer a distraction. One significant benefit of CBT is the availability of instant results. You can receive your test results before you leave the test center. CBT testing offers an increased degree of flexibility in scheduling. Testing costs for CBT tests are slightly higher than paper-and-pencil tests and the number of testing sites is limited. First-time test takers may be more comfortable with the paper-and-pencil tests but each technician has an option.

Who Writes the Questions?

Service industry experts in the area being tested write the questions. Each area has its own technical experts. Questions are entirely job related, and are designed to test the skills you need to be a successful technician. Theoretical knowledge is important and necessary to answer the questions but ability to apply that knowledge is the basis of ASE test questions.

Each question has its roots in an ASE "item-writing" workshop where service representatives from automobile manufacturers (domestic and import), aftermarket parts and equipment manufacturers, working technicians, and vocational educators meet to share ideas and translate them into test questions. Each test question written by these experts must survive review by all members of the group. Questions are written to address practical application of soft skills and system knowledge experienced by technicians in their day-to-day work.

All questions are pretested and quality-checked on a national sample of technicians. Questions that meet ASE standards of quality and accuracy are included in the scored sections of the tests; the "rejects" are sent back to the drawing board or discarded altogether.

Each certification test is made up of 40–80 multiple-choice questions.

Note: Each test could contain additional questions that are included for statistical research purposes only. Your answers to these questions will not affect your score, but since you do not know which ones they are, you should answer all questions in the test. The 5-year Recertification Test covers the same content areas as those listed above. However, the number of questions in each content area of the Recertification Test will be reduced by about one half.

Objective Tests

A test is called an objective test if the same standards and conditions apply to everyone taking the test and there is only one correct answer to each question.

Objective tests primarily measure your ability to recall information. A well-designed objective test can also test your ability to understand, analyze, interpret, and apply your knowledge. Objective tests include true-false, multiple choice, fill in the blank, and matching questions. ASE's tests consist exclusively of 4-part multiple-choice objective questions.

The following are some strategies that may be applied to your tests.

Before beginning to take an objective test, quickly look over the test to determine the number of questions, but do not try to read through all of the questions. In an ASE test there are usually 40—80 questions, depending on the subject. Read through each question before marking your answer. Answer the questions in the order they appear on the test. Leave questions that you are not sure of blank and move on to the next question. You can return to the unanswered questions after you have finished the others. They may be easier to answer after your mind has had additional time to consider them on a subconscious level. In addition, you might find information in other questions that will help you recall the answers to some of them.

Do not be obsessed by the apparent pattern of responses. For example, do not be influenced by a pattern like **D, C, B, A, D, C, B, A** on an ASE test.

There is also a lot of folk wisdom about taking objective tests. For example, some people advise that you avoid response options that use certain words such as *all, none, always, never, must,* and *only,* to name a few. This, they claim, is because nothing in life is exclusive. They would advise that you choose response options that use words that allow for some exception, such as *sometimes, frequently, rarely, often, usually, seldom,* and *normally.* Some people would also advise that you avoid the first and last option (A and D) because they feel test writers are more comfortable if they put the correct answer in the middle (B and C) of the choices. Another recommendation often offered is to select the option that is either shorter or longer than the other three choices because it is more likely to be correct. Some would advise that you never change an answer since your first intuition is usually correct.

Although there may be a grain of truth in this folk wisdom, it is not wise to make your selections based on the length or order of the answer choices. ASE test writers try to ensure that there are just as many **A** answers as there are **B** answers, just as many **D** answers as **C** answers. As a matter of fact, ASE tries to balance the answers at about 25 percent per choice **A, B, C,** and **D**. There is no intention to use "tricky" words, such as those outlined above. Put no credence in the opposing words "sometimes" and "never," for example.

Multiple-choice tests are sometimes challenging because there are often several choices that may seem possible, and it may be difficult to decide on the correct one. The best strategy, in this case, is to first determine the correct answer before looking at the options. If you see the answer you decided on, you should still examine the options to make sure that none seem more correct than yours. If you do not know or are not sure of the answer, read each option very carefully and try to eliminate those options that you know to be wrong. That way, you can often arrive at the correct choice through a process of elimination.

If you have gone through all of the test and you still do not know the answer to some of the questions, *then guess*. Yes, guess. You then have at least a 25 percent chance of being correct. If you leave the question blank, you have no chance. Your score is based on the number of questions answered correctly.

Preparing for the Exam

We have included many sample and practice questions in this guide simply to help you learn what you know and what you do not know. We recommend that you work your way through each question in this book. Before doing this, carefully look through Section 3; it contains a description and explanation of the type of questions you'll find in an ASE exam.

Once you understand what the questions will look like, move to the sample test. Answer one of the sample questions (Section 5) then read the explanation (Section 7) to the answer for that question. If you don't feel you understand the reasoning for the correct answer, go back and read the overview (Section 4) for the task that is related to that question. If you still don't feel you have a solid understanding of the material, identify a good source of information on the topic, such as a textbook, and do some more studying.

After you have completed all of the sample test items and reviewed your answers, move to the additional questions (Section 6). This time answer the questions as if you were taking the actual test. Do not use any reference or allow any interruptions so that you can get a feel for how you will do on an actual test. Once you have answered all of the questions, grade your results using the answer key in Section 7. For every question that you answered incorrectly, study the explanations to the answers and/ or the overview of the related task areas. Try to determine the cause for missing the answer. The easiest thing to correct is learning the correct technical content. The hardest things to correct are behaviors that lead you to a wrong conclusion. If you knew the information but still got it wrong, there is a behavior problem that will need to be corrected. For example, you may be reading too quickly and skipping over words that affect your reasoning. If you can identify what you did that caused you to answer the question incorrectly, you can eliminate that cause and improve your score. Here are some basic guidelines to follow while preparing for the exam:

- Focus your studies on those areas in which you are weak.

- Be honest with yourself while determining if you understand something.

- Study often but in short periods of time.

- Remove yourself from all distractions while studying.

- Keep in mind the goal of studying is not just to pass the exam, the real goal is to learn!

- Prepare physically by getting a good night's rest before the test and eat meals that provide energy but do not cause discomfort.

- Arrive early to the test site to avoid long waits as test candidates check in and to allow all of the time available for your tests.

During the Test

On paper-and-pencil tests you will be placing your answers on a sheet where you will be required to blacken (bubble) in your answer choice. Stray marks or incomplete erasures may be picked up as an answer by the electronic reader so be sure only your answers end up on the sheet. One of the biggest problems an adult faces in test taking is in placing an answer in the correct spot on a bubble sheet. Make certain that you mark your answer for, say, question 21, in the space on the bubble sheet designated for the answer for question 21. A correct response in the wrong line will probably result in two questions being marked wrong, one with two answers (which could include a correct answer but will be scored wrong), and the other with no answer. Remember, the answer sheet on written test is machine scored and can only "read" what you have bubbled in.

If you finish answering all of the questions on a test and have remaining time, go back and review the answers to those questions that you were not sure of. You can often catch careless errors by using the remaining time to review your answers. Carefully check your answer sheet for blank answer blocks or missing information.

At practically every test, some technicians will invariably finish ahead of time and turn in their papers long before the final call. Some technicians may be doing recertification tests and others may be taking fewer tests than you. Do not let them distract or intimidate you.

It is not wise to use less than the total amount of time that you are allotted for a test. If there are any doubts, take the time for review. Any product can usually be made better with some additional effort. A test is no exception. It is not necessary to turn in your test paper until you are told to do so.

Testing Time Length

An ASE written test session is 4 hours. You may attempt from 1 to a maximum of 4 tests in one session. It is recommended, however that no more than a total of 225 questions be attempted at any test session. This will allow for just over 1 minute for each question.

Visitors are not permitted at any time. If you want to leave the test room, for any reason, you must first ask permission. If you finish your test early and want to leave, you are permitted to do so only during specified dismissal periods. You are not permitted back in the test room once you leave after completing your test.

You should monitor your progress and set an arbitrary limit as to how much time you will need for each question. This should be based on the number of questions you are attempting. Wear a watch because some facilities may not have a clock visible to all areas of the room.

CBTs are allotted a testing time according to the number of questions ranging from 1/2 hour to 1 1/2 hours. Advanced-level tests are allowed 2 hours. This time is by appointment so arrive on time to ensure that you have all of the time allocated. If you arrive late for a CBT test appointment, you will only have the amount of time remaining on your appointment.

Your Test Results!

You can gain a better perspective about tests if you know and understand how they are scored. ASE's tests are scored by American College Testing (ACT) service, a nonpartial, unbiased organization having no vested interest in ASE or in the automotive industry.

Each question carries the same weight as any other question. For example, if there are 50 questions, each is worth 2 percent of the total score. The passing grade is 70 percent. That means you must correctly answer 35 of the 50 questions to pass the test.

The test results can tell you:

- where your knowledge equals or exceeds that needed for competent performance, or
- where you might need more preparation.

Your ASE test score report is divided into content areas and will show the number of questions in each content area and how many of your answers were correct. These numbers provide information about your performance in each area of the test. However, because there may be a different number of questions in each content area of the test, a high percentage of correct answers in an area with few questions may not offset a low percentage in an area with many questions.

One does not "fail" an ASE test. The technician who does not pass is simply told "More Preparation Needed." Though large differences in percentages may indicate problem areas, it is important to consider how many questions were asked in each area. Since each test evaluates all phases of the work involved in a service specialty, you should be prepared in each area. A low score in one area could keep you from passing an entire test. There is no such score as average. You cannot determine your overall test score by adding the percentages given for each task area and dividing by the number of areas. Test scoring does not work that way because generally there are not the same number of questions in each task area. A task area with 20 questions, for example, counts more toward your total score than a task area with 10 questions.

Your test report should give you a good picture of your results and a better understanding of your strength and weaknesses for each task area.

If you do not pass the test, you may take it again at any time it is scheduled to be administered. You are the only one who will receive your test score. Test scores will not be given over the telephone by ASE nor will they be released to anyone without your written permission.

3 | Types of Questions on an ASE Exam

ASE certification tests are often thought of as being tricky. They may seem tricky if you do not completely understand what is being asked. The following examples will help you recognize certain types of ASE questions and avoid common errors.

Paper-and-pencil tests and Computer Based Test questions are identical in content and difficulty. Most initial certification tests are made up of 40–80 multiple-choice questions. Multiple-choice questions are an efficient way to test knowledge. To answer them correctly, you must think about each choice as a possibility, and then choose the one that best answers the question. To do this, read each word of the question carefully. Do not assume you know what the question is about until you have finished reading it.

About 10 percent of the questions on an actual ASE exam will use an illustration. These drawings contain the information needed to answer the question correctly. The illustration must be studied carefully before attempting to answer the question. Often, technicians look at the possible answers then try to match up the answers with the drawing. Always do the opposite; match the drawing to the answers. When the illustration is showing an electrical schematic or another system in detail, look over the system and try to figure out how the system works before you look at the question and the possible answers.

Multiple-Choice Questions

The most common type of question used on ASE Tests is the multiple-choice test. This type of question contains 3 "distracters" (wrong answers) and 1 "key" (correct answer). When the questions are written, effort is made to make the distracters plausible to draw an inexperienced technician to one of them. This type of question gives a clear indication of the technician's knowledge. Using multiple criteria including cross-sections by age, race, and other background information, ASE is able to guarantee that a question does not bias for or against any particular group. A question that shows bias toward any particular group is discarded.

If you encounter a question that you are unsure of, reverse engineer it by eliminating the items that it cannot be. For example:

Under load, a diesel engine emits dark black smoke from the exhaust pipe. What should the technician check first?

A. air cleaner

B. fuel pump

C. injector pump

D. turbocharger (A4)

Answer A is correct. Whenever a diesel engine emits black smoke, oxygen starvation caused by air cleaner restriction should be checked first, mainly because it can be checked and quickly eliminated as a cause. Test the intake air inlet restriction using a water manometer. The specifications should always be checked to the OEM values, but typical maximum values will be close to:

1. 15 in. H_2O vacuum—naturally aspirated engines

2. 25 in. H_2O vacuum—boosted engines

Answer B is incorrect. The fuel pump (transfer pump) is very unlikely to cause a black smoke condition.

Answer C is incorrect. The injector pump may cause black smoke, but it is not the first item to check.

Answer D is incorrect. The turbocharger can cause black smoke, but it is not the first item to check.

EXCEPT Questions

Another type of question used on ASE tests has answers that are all correct except one. The correct answer for this type of question is the answer that is wrong. The word "EXCEPT" will always be in capital letters. You must identify which of the choices is the wrong answer. If you read quickly through the question, you may overlook what the question is asking and answer the question with the first correct statement. This will make your answer wrong. An example of this type of question and the analysis is as follows:

All of the following are steps in an air inlet restriction test EXCEPT:

A. Connect a manometer to the air intake downstream from the filter.

B. Remove the air cleaner element and duct.

C. Check the manometer reading with the engine under load.

D. Record the inlet restriction spec in inches of H_2O.   (A7)

Answer B is correct. The steps outlined here to test air intake/inlet restriction are correct with the exception of removing the air cleaner element as the main reason for performing this test to verify the serviceability of the filter element. To perform an air inlet restriction test:

• Connect a vacuum gauge, water manometer, or Magnehelic gauge to the intake air piping according to the OEM's instructions.

• Start the engine and operate at the OEM-required load and speed.

• Measure the pressure drop at the turbocharger inlet and record the reading. The typical maximum reading is approximately 25 in. H_2O (635 mm H_2O).

Answer A is incorrect. Connecting a manometer is a step.

Answer C is incorrect. The engine should be under load.

Answer D is incorrect. The reading is in inches of water.

Technician A, Technician B Questions

The type of question that is most popularly associated with an ASE test is the "Technician A says . . . Technician B says . . . Who is correct?" type. In this type of question, you must identify the correct statement or statements. To answer this type of question correctly, carefully read each technician's statement and judge it on its own merit to determine if the statement is true.

Sometimes this type of question begins with a statement about some analysis or repair procedure. This is often referred to as the stem of the question and provides the setup or background information required to understand the conditions on which the question is based. This is followed by two statements about the cause of the concern, proper inspection, identification, or repair choices. You are asked whether the first statement, the second statement, both statements, or neither statement is correct. Analyzing this type of question is a little easier than the other types because there are only two ideas to consider although there are still four choices for an answer.

Technician A, Technician B questions are really double true or false questions. The best way to analyze this kind of question is to consider each technician's statement separately. Ask yourself, is A true or false? Is B true or false? Then select your answer from the four choices. Remember, an ASE Technician A, Technician B question will never have Technician A and B directly disagreeing with each other. That is why you must evaluate each statement independently.

An example of this type of question and the answer for it follows.

A set of dry liners is to be fitted to a diesel engine cylinder block. Technician A says that it is good practice to fit the liners selectively to block by fitting the liner with the largest OD to the block bore with the largest ID and progressing downward. Technician B says that it is important to machine the correct cross-hatch pattern into the block bore before fitting each liner. Who is correct?

A. A only

B. B only

C. Both A and B

D. Neither A nor B   (C3)

Answer A is correct. Only Technician A is correct. Technician A correctly describes the process of selectively fitting a set of dry liners to a cylinder block, a recommended practice.

Answer B is incorrect. Cross-hatch is not required to be cut to the cylinder block bore in any engine that uses liners; this is machined to the liner bore. The dry liner uses thinner walled sleeves than the wet liner. Dry liners (sleeves) are installed into the block bore, usually with a marginally loose fit, and are retained by the cylinder head. Dry sleeves do not transfer heat as efficiently as wet liners, but they are easily replaced and do not present coolant-sealing problems.

Answer C is incorrect. Only Technician A is correct.

Answer D is incorrect. Only Technician A is correct.

Most-Likely Questions

Most-Likely questions are somewhat difficult because only one choice is correct while the other three choices are nearly correct. An example of a Most-Likely-cause question is as follows:

An electric oil pressure gauge indicates 0 pressure, but the engine is well lubricated. Which of the following is Most-Likely at fault?

A. The oil passage to the sensor is open.

B. The wrong engine oil was installed.

C. The wiring to the sensor is open.

D. The engine is not at operating temperature.   (D1)

Answer C is correct. Because the question tells you that the engine is operating well and the gauge reading is 0, the cause of the problem is Most-Likely an electrical open condition in the gauge/ sensor circuit. The other answers describe conditions that are unlikely to result in a 0 reading although they would produce inaccurate/erratic readings. A piezoresistive sensor is the type of sender often used for oil pressure gauges. An ohmmeter is used to check this type of sender by connecting the leads to the sending unit terminal and ground. First, check the resistance with the engine off and compare to specifications. Next, start the engine and allow it to idle. Recheck the resistance value and compare it to specifications. Do not condemn this electrical sending unit immediately if the readings are not within specifications. First, connect a mechanical pressure gauge to the engine to confirm that it is producing adequate oil pressure.

Answer A is incorrect. The oil passage should be open.

Answer B is incorrect. The wrong oil most likely would not cause the gauge to read 0. It may read low but not 0.

Answer D is incorrect. If the engine was not at operating temperature, the oil pressure would most likely be higher than normal, causing a high gauge reading, not 0 pressure.

LEAST-Likely Questions

Notice that in Most-Likely questions not all the letters are capitalized. This is not so with LEAST-Likely questions. For this type of question, look for the choice that would be the LEAST-Likely cause of the described situation. Read the entire question carefully before choosing your answer. Avoid relating questions to those unusual situations that you may have encountered and answer based on the technical and mechanical possibilities.

An example is as follows:

When inspecting an oil filter housing or its mounting, which of the following is LEAST-Likely to be a procedure?

A. visually check for cracks

B. visually check gasket surface for nicks

C. magnaflux the housing to detect small cracks

D. visually inspect housing passageways for obstructions   (D3)

Answer C is correct. The LEAST-Likely part of the procedure is checking a filter pad housing for cracks with magnetic flux equipment. Visual checking is usually all that is required to check out a filter pad properly.

Answer A is incorrect. The housing should be visually checked for cracks.

Answer B is incorrect. The gasket surfaces should be checked for nicks.

Answer D is incorrect. The passages should be inspected for obstructions.

Summary

There are no four-part multiple-choice ASE questions having "none of the above" or "all of the above" choices. ASE does not use other types of questions, such as fill-in-the blank, completion, true-false, word-matching, or essay. ASE does not require you to draw diagrams or sketches. If a formula or chart is required to answer a question, it is provided for you. There are no ASE questions that require you to use a pocket calculator.

4 Overview of Task List

Diesel Engines (Test H2)

The following section includes the task areas and task lists for this test and a written overview of the topics covered in the test.

The task list describes the actual work you should be able to do as a technician and that you will be tested on by the ASE. This is your key to the test; you should review this section carefully. The sample test and additional questions are based on these tasks, and the overview section will also support your understanding of the task list.

ASE advises that the questions on the test may not equal the number of tasks listed; the task lists indicate what ASE expects you to know how to do and be ready to be tested on.

At the end of each question in the Sample Test and Additional Test questions sections, a letter and number are used as a reference back to this section for additional study. Note the following example: A15.

A. General Engine Diagnosis (17 Questions)

Task A15   **Check lubrication system for contamination, oil level, quality, temperature, pressure, filtration, and oil consumption; determine needed repairs.**

Example:

1. Technician A says that to verify the readings on a dash oil pressure gauge, they should be identical to diagnostic master gauge readings. Technician B says test readings should be taken during startup, varying operating ranges, and on shutdown. Who is correct?
   A. A only
   B. B only
   C. Both A and B
   D. Neither A nor B   (A15)

Question #1

Answer C is correct. Both technicians are correct. Readings on a dash oil pressure gauge should be identical to diagnostic master gauge readings. The readings should be taken during startup, at varying operating ranges, and on shutdown.

Answer A is incorrect. Technician B is also correct.

Answer B is incorrect. Technician A is also correct.

Answer D is incorrect. Both technicians are correct.

Task List and Overview

A. General Engine Diagnosis (17 Questions)

Task A1   **Verify the complaint, and road test vehicle; review operator service request and past maintenance documents (if available); determine further diagnosis.**

The service technician must be aware of normal diesel engine operating noises to determine whether a system requires service. Normal noises include the sounds of diesel combustion, fuel injectors, etc. Abnormal noise includes sounds that are produced by misfire, diesel knock, valve clatter, vibration, etc.

The diesel engine technician needs to apply a step-by-step diagnostic procedure to determine the cause of the problem. The following is a typical diagnostic routine:

- Listen carefully to the operator's complaint.

- Road test the bus if necessary to identify the complaint.

- Perform diagnostic tests to locate the root cause of the complaint. Start with the easiest, quickest tests, and then work toward the more difficult and time-consuming tests.

- After performing the required repairs, be sure the operator's complaint is eliminated; road test the bus to verify the repair that eliminated the cause.

Task A2   **Inspect engine assembly and engine compartment for fuel, oil, coolant, exhaust, or other leaks; determine needed repairs.**

To reduce noxious emissions, engine manufacturers design engines to reduce venting gases and fumes. These efforts have also produced a cleaner engine and engine compartment because fluid leakage is less likely. Careful inspection of the engine and engine compartment can reveal potential problems. For example, oil residue on the inside of the engine compartment door indicates excess crankcase pressure caused by blowby or an overfilled crankcase. Oil on the exhaust joints could indicate a leaking head gasket or a leaking turbocharger oil seal.

Engine stains that appear discolored or lighter than the surrounding area could be caused by a small coolant leak from a hose or the head. A light, slippery, almost invisible coating on most of the engine compartment surfaces could be caused by a small leak in the radiator. This coating could be from excess venting of radiator pressure or a very small leak in a high-pressure fuel or hydraulic line.

Deposits of liquid in cracks and crevices or low spots can be indicators of leaks from the surrounding components, lines, and/or hoses. Leak detection is a common-sense skill. Remember, gravity causes liquids to accumulate at the lowest point. If fluids are lost, they usually flow downward or out of the exhaust.

A leak detection method favored by some technicians is to add a special dye to the fluid that is leaking. The bus is operated long enough for the leak to become apparent. A special light is used that causes the dye to appear fluorescent, and can easily be traced back to the source of the leak. Dyes are available for addition to fuel, coolant, oils, and refrigerants.

Task A3   **Inspect engine compartment wiring harness, connectors, seals, and locks; check for proper routing and terminal/connector condition; determine needed repairs.**

Modern engine electrical/electronic wiring is complex. If connectors or wires fail, the circuit sends faulty inputs from the monitoring sensors that operate on milliamps of current. This, in turn, causes computerized circuits using these inputs to make inappropriate adjustments to the engine control. Some common problems associated with circuits are corrosion and other foreign substances in connectors, loose or improperly mated connectors, and improper grounding. Some of these problems will not affect engine operation until several other problems exist, by which time, they can complicate the troubleshooting process. If a visual inspection of a connector reveals corrosion, the connector should be cleaned or replaced. Wiring that is visibly lumpy or has nicked insulation should be repaired or replaced to prevent faults from corrosion. Loose or scale-coated grounds should be removed, cleaned, treated for corrosion, reinstalled, and coated with a corrosion inhibitor. All connections to a ground must be free of rust and/or paint to ensure a good electrical connection.

Task A4   **Listen for and diagnose engine noises; determine needed repairs.**

A technician who develops a discriminating ear for engine noises can simplify troubleshooting or find problems before they cause breakdowns. Listening through a metal bar or heater hose applied to each cylinder in sequence can expose a misfiring cylinder like a stethoscope. Sometimes the absence of change in the sound may provide confirmation of a faulty condition. Again, if one of the cylinders appears to be misfiring, short out each injector mechanically or by performing an electronic cylinder cut-out test. Every time you short out a good cylinder, the sound of the engine should change. On the other hand, shorting out a dead cylinder will provide no noticeable change in sound. Often the frequency of an unusual noise is relevant to the defective component's operating speed. A loud knocking that increases at the same rate as the engine would probably be associated with the crank. Whereas a sound that is half the engine speed might be created by a cam or something directly attached to it. Other components may produce high-frequency sounds like a squeal, and may not be as discernible using this method. In these cases, turning your head from side to side to determine the direction of loudest sound may point to the general location. Then, look for components in that area that operate at a high speed. Squeals may also be produced when two surfaces are rubbing each other, like belts slipping. Sometimes sounds appear random in nature. Usually the operator, an automated operation, or a load change, like a power steering pump squealing when the steering wheel is turned, causes these randomly occurring noises.

Task A5   **Check engine exhaust emissions, odor, smoke color, opacity (density), and quantity; determine needed repairs.**

The color and consistency of the exhaust can reveal problems associated with the combustion of the engine. When a diesel engine is running properly with a load, its exhaust should be clear. Combustion problems produce discoloration or smoke. Variations in engine operation, such as changes in fuel demand, loading, and outside temperature can produce a small degree of smoke. Excessive or prolonged appearance of smoke is an indicator of other problems. Oil, for instance, adds a blue haze to the exhaust, causing a blue smoke. Unburned fuel or an improper grade of fuel produces a gray or black smoke; if this becomes excessive, soot will form. This condition could also occur if the intake system has provided insufficient air for fuel delivered. Water or coolant will produce a white smoke-like steam. White smoke may also indicate that injectors are misfiring; this happens when combustion does not occur and fuel vapors are present in the exhaust. White smoke occurs either when the engine starts the first time or when the ambient temperature is too low for proper combustion.

Because of incomplete mixing of the injected fuel with the air in the combustion chamber, some fuel droplets do not burn until late in the cycle. This means that the combustion chamber temperature is lower and there is also less oxygen to sustain the remaining burn; therefore, incomplete combustion occurs. This situation results in smoke coming out the exhaust. White smoke results from incomplete combustion in overlean combustion chamber areas, by fuel spray condensation on metal surfaces, and with low temperatures in the cylinder, such as when starting an engine (more so on cold days). Fuel with too low a cetane rating can also cause white smoke when used in low ambient-temperature conditions. White smoke occurs more often in an indirect injected (IDI) engine from retarded timing. Late timing on a direct injected (DI) engine also causes white smoke. Gray or black smoke is the result of incomplete combustion in rich combustion chamber areas, caused by such conditions as engine overload, insufficient fuel injector spray penetration, late ignition due to retarded injection timing, or poor fuel evaporation and mixing due to advanced timing. Air starvation is the major cause of black smoke.

Note on MIXING: Owing to the extremely short time available for mixing, as the fuel-air rate increases, an appreciable fraction of the fuel fails to find the necessary oxygen (O_2) for combustion and passes through the cylinder unburned or partially burned. This condition is more common in older engines.

Task A6   **Perform fuel supply and fuel return system tests; check fuel for contamination, quality/type/grade, and consumption; determine needed repairs.**

If the engine runs but is low on power, check the following items to identify the problem:

- Check OEM service literature before performing tests.
- Check the fuel filter and change it if necessary.
- Check the throttle arm travel on mechanically controlled engines. If the throttle travel is not correct, check to make sure the throttle is in the wide-open position with the accelerator pedal at full travel.

Fuel starvation will generally result in total combustion of fuel supplied with a loss of power, no noticeable change in exhaust color, and low cylinder temperatures. Some of the reasons for fuel starvation are an empty tank, defective fuel pump, leaks in fuel lines from the tank, restricted filters or strainers, and crimped fuel lines between the pump and injectors. A temperature spread of less than 10°F (6°C) between the fuel's pour point and temperature can restrict fuel flow enough to cause starvation. Faulty injectors or improper spray patterns can also reduce power and cause erratic operation; this is usually accompanied by black or gray smoke.

Gray or black smoke is the result of incomplete combustion in rich combustion chamber areas, caused by engine overload, insufficient fuel injector spray penetration, late ignition due to retarded injection timing, or poor fuel evaporation and mixing due to advanced timing.

Task A7   **Perform air intake system restriction and leakage tests; determine needed repairs.**

When an engine suffers from a loss of power or rough operation, you should go back to the basics. To obtain the optimum performance from an engine, the proper quantity and quality of fuel must be supplied to the cylinder at the right time with the right amount of air and heat.

Air starvation will generally result in incomplete combustion, increasing black smoke or soot, a decrease in power, and heat build-up in the cylinder due to insufficient scavenging. Some of the reasons for air starvation are restricted intake hoses and pipes, dirty air filter, excessive exhaust back pressure, or high altitude operation. When the diesel engine is equipped with a blower or turbocharger, check for leaks in the output tubes and the aftercooler. Verify that the aftercooler is functioning properly. Ultimately, the denser the air, the better the engine runs. Air starvation is the major cause of black smoke. Always check for an inlet restriction problem first when investigating a black smoke complaint.

Air inlet restriction may be measured in a number of ways. Most buses are equipped with resettable air inlet restriction gauges that are either mounted directly to the filter canister, or remotely mounted at the dash. They are not as accurate as test equipment, but do provide an indication of when the element needs to be changed. Inlet restriction is measured in inches of water vacuum. Most manufacturers recommend filter element replacement at 25 inches of water vacuum. To obtain accurate measurements with test equipment, such as a water manometer or a magnehelic gauge, the engine should be under full load, and the gauge or manometer should be connected just downstream from the filter. When it is not possible to place the engine under full load, a snap throttle test will give a close approximation. Visually check the intake pipe for evidence of dirt entry and make immediate repairs if any is found, since dirt can quickly destroy the engine. Low power complaints under load may also result from a rubber hose from the air cleaner that collapses, and starves the engine for air.

Task A8   **Perform intake manifold pressure tests; determine needed repairs.**

Most current diesel engines are turbocharged so the intake manifold pressure produced by the turbocharger is critical for optimum engine performance. Turbocharging compresses the air charge to the engine cylinder making it denser—that is, more oxygen rich. Compressing air heats it to temperatures exceeding 300°F. Downstream from the turbocharger, a heat exchanger is located to help cool the air charge before it is delivered to the engine cylinder. Several types are used but the most common uses ambient air drawn across transfer tubes, transferring the heat from the intake air to the ambient air. These are known as charge air coolers. Another type uses engine coolant routed through a core to help cool boost air.

Small leaks in the either piping downstream from the turbocharger or charge air cooler itself can lower the manifold boost pressure and cause low power. Peak manifold pressure can only be measured when the engine is under full load, so actual manifold boost is best tested to specification when a chassis dyno is available. Stall testing procedures can be used to check for maximum boost pressure, but care must be exercised not to overload drive train components. Leaks can be identified using a soapy water solution. Small leaks in a charge air cooler can usually be located by plugging the inlet and outlets and pressurizing with regulated air pressure to a value of around 5 psi.

Task A9   **Perform exhaust back-pressure and temperature tests; determine needed repairs.**

A slight pressure in the exhaust system is normal. However, excessive exhaust back pressure seriously affects engine operation. In 2-stroke cycle engines, it may cause an increase in the air box pressure with a resultant loss of efficiency of the blower. This means less air for scavenging, which results in poor combustion and higher temperatures. High exhaust back pressure causes a loss of performance due to poor combustion and higher exhaust temperatures. Causes of high exhaust back pressure are usually a result of an inadequate or improper type of muffler, an exhaust pipe that is too long or too small in diameter, an excessive number of sharp bends in the exhaust system, or obstructions such as excessive carbon formation or foreign matter in the exhaust system. Newer vehicles equipped with Diesel Oxidation Catalysts (DOCs) or Diesel Particulate Traps (DPFs) may have electronic monitoring of back pressure and temperature. DOCs and DPFs are subject to clogging and overheating, and should be monitored for obvious safety reasons. Excessive fuel or engine oil present in the exhaust can cause clogging and overheating problems. Check the exhaust back pressure, measured in inches of mercury, with a manometer. Connect the manometer to the exhaust manifold (except on turbocharged engines) by removing a pipe plug which is provided for that purpose. If no opening is provided, drill a 11/32-inch hole in the exhaust manifold companion flange and tap the hole to accommodate a pipe plug. On turbocharged engines, check the exhaust back pressure in the exhaust piping 6 to 12 inches from the turbine outlet. The tapped hole must be in a comparatively straight pipe area for an accurate measurement.

Task A10   **Perform crankcase pressure test; determine needed repairs.**

The crankcase pressure indicates the amount of cylinder gas passing by the piston rings into the crankcase. A slight pressure in the crankcase is desirable to prevent the entrance of dust. A loss of engine lubricating oil through the breather tube, crankcase ventilator, or dipstick hole in the cylinder block is indicative of excessive crankcase pressure. The causes of high crankcase pressure may be traced to excessive blowby due to worn piston rings, a hole or crack in a piston crown, loose piston pin retainers, worn blower oil seals, defective blower, cylinder head gaskets, or excessive exhaust back pressure. In addition, the breather tube or crankcase ventilator should be checked for obstructions. Check the crankcase pressure with a manometer connected to the oil-level dipstick opening in the cylinder block. Check the readings obtained at various engine speeds with the manufacturer's manual.

Task A11   Diagnose no cranking, cranks but fails to start, hard starting, and starts but does not continue to run problems; determine needed repairs.

When a diesel engine cranks but does not start, check that the engine is being fueled by observing the exhaust. In most current diesel engines, fueling is controlled by an engine management computer so at the first stage of troubleshooting it will be necessary to consult the self-diagnostics built into the system. In most cases, this will mean that the manufacturer's diagnostic literature, electronic service tools, and some specialized training are required. Never attempt to trial-and-error troubleshoot electronically controlled engines because costly damage and downtime can result.

When an engine cranks but fails to start, often this indicates air in the fuel system. To determine if air is causing the problems, pressurize the system using a hand primer and bleed the lines to each injector while cranking the engine. If the primer pump fails to build up pressure within the required time, check fuel filter lines for leaks and the tank for adequate fuel level. Observe the return fuel flow for the presence of air. If air is found, check the suction side of the fuel system for potential leaks.

A diesel engine may not start or be hard to start because the temperature of the air in the cylinder is too low for full combustion of the atomized fuel. If the starter does not rotate the engine fast enough to allow the air in the cylinders to reach combustion temperature, combustion will not occur.

Electrically, the starter may turn slowly if it is defective, has a low battery potential, has corroded cables, or corroded electrical connectors. Inspect the starter, battery, and connections to ensure proper operation of the starter motor.

Mechanically, the engine may offer too large a load on the starter. This can occur when the oil is too thick or the surfaces too cold, the belt or gear driven devices are loading down the engine, or there is too much cylinder back pressure. Check oil specifications for that climate zone, check engine preheaters if installed, check for proper belt tension and gear backlash, and use unloader valves if installed.

- Low combustion temperature may also result if the outside temperature is too low and you do not use preheating.

- To assist in starting the engine at low temperatures use ether to reduce the flash point of the fuel/air mixture.

- To maintain a warm engine for starting at low temperatures the engine may use an immersion heater in the cooling system or submerged in the crankcase oil.

Reduced compression pressure created by a defective turbo or blower, a restricted air intake or excessive exhaust back pressure will also delay or prevent combustion pressure build-up. Engines with electronic fuel injection systems may not start if vital sensor data is not received.

Task A12   Diagnose surging, rough operation, misfiring, low power, slow deceleration, slow acceleration, and shutdown problems; determine needed repairs.

When an engine suffers from a loss of power or rough operation, you should go back to the basics. To obtain the optimum performance from an engine, the proper quantity and quality of fuel must be supplied to the cylinder at the right time with the right amount of air and heat.

Air starvation will generally result in incomplete combustion, increasing black smoke or soot, a decrease in power, and heat build-up in the cylinder due to insufficient scavenging. Some reasons for air starvation are restricted intake hoses and pipes, dirty air filter, excessive exhaust back pressure, and high altitude operation. If the diesel comes equipped with a blower or turbocharger, check for leaks in the output piping and the aftercooler. Verify that the aftercooler is functioning properly. Ultimately, the denser the air, the better the engine runs. Fuel starvation will generally result in total combustion of fuel supplied with a loss of power, no noticeable change in exhaust color, and low cylinder temperatures. Some reasons for fuel starvation are an empty tank, defective fuel pump, leaks in fuel lines from the tank, restricted filters or strainers, and crimped fuel lines between the pump and injectors. A temperature spread of less than 10°F (6°C) between the fuel's pour point and ambient temperature can restrict fuel flow enough to cause starvation. Faulty injectors or improper spray patterns can also reduce power and cause erratic operation; this is usually accompanied with black or gray smoke.

Misfiring cylinders may be located in several ways. One method is to compare exhaust temperatures of each cylinder. They should be uniform. When a cylinder is located that has notably cooler exhaust temperature, it is most likely misfiring. A variety of equipment is available to measure exhaust temperature of individual cylinders, including temperature probes, and no contact, infrared "heat guns." Other ways of isolating misfiring cylinders depend on the type of fuel system used. With pump-line-nozzle fuel systems, it is usually possible to open each injector line momentarily, "shorting out" that cylinder, while observing engine rpm, idle quality, and the color of the exhaust. A cylinder that does not make significant difference is misfiring. Engines with mechanical unit injectors may require substituting a known good injector or swapping with another injector from a cylinder that is not misfiring to determine whether the injector is at fault, or if there is a mechanical problem such as a worn cam follower, or low compression. Engines with electronically managed fuel systems often provide for dynamic cylinder balance testing. Follow the software instructions and prompts to perform these tests.

Surging is an uneven power output from the engine that is usually the result of either fuel starvation, or small quantities of air in the fuel. Air can enter the system through hoses, fittings, filters, and filter housings on the supply side of the transfer (lift) pump. Air entry may also be the result of a low fuel level in the tank, or a standpipe that is cracked or not sufficiently submerged. To check for the presence or air, a sight glass may be used on either the supply side of the system or on the fuel return line.

Slow deceleration problems can be caused by a restricted fuel return line. Check the flow and temperature of the fuel return for an adequate flow of warm fuel.

Task A13   Isolate and diagnose engine related vibration problems; check engine mounts; determine needed repairs.

Often the frequency of occurrence of an unusual vibration is a direct or multiple of the operational speed of the device creating the vibration. A strong vibration that increases in occurrence rate with engine speed would probably be the crank or something directly attached to it. Whereas a vibration that has an occurrence rate half the engine speed might be created by a cam or something directly attached to it. Other components, like an out-of-balance turbocharger or alternator, may produce high-frequency vibrations that seem to have a constant occurrence rate. In these cases, use a stethoscope to find the strongest concentration of that vibration. Then, look for components in that area that operate at a high speed. High-frequency vibrations can also be produced by bearings that are failing or two surfaces that are rubbing each other. These are usually destructive in nature, have a constant occurrence rate, progressively grow in intensity, and last for a short time before they become more apparent. Some automated operations, like a cycling air-conditioning compressor, may have a vibration occurrence rate that seems random. Two or more vibrations of different frequency at the same time may produce a rhythmic vibration.

Task A14   Check cooling system for temperature protection level, contamination, coolant type and level, temperature, pressure, supplemental coolant additive (SCA) concentration, filtration, and fan operation; determine needed repairs.

The radiator cap will maintain the pressure within the cooling system at a specified value, usually 15 psi (103 kPa). During heavy loads and prolonged periods of idling, this pressure may exceed the rating of the radiator cap and fluid will be vented to an expansion tank or bottle. Once enough fluid has been vented to reduce the pressure within limits, the spring tension of the radiator cap will close the vent line. At normal temperatures, with the pressure cap removed, small fluctuations in coolant level can be expected. Inspection of this fluid level may indicate deeper problems. A constant overflowing of the fluid from the tube may indicate air pockets, uneven heating, or hot spots in the engine. Repetitive and rhythmic excessive surges in this level usually indicate a leak to the combustion cylinder, such as a blown head gasket.

At different and spontaneous times, the cooling system will overheat. At the time, the system should be checked as a unit in this manner: check tightness on hose clamps, check for soft or sometimes mushy feeling hoses, and check for cracks. You should be able to see all the components. Look for leaks, such as at the radiator, water pump, cooling filter, head gaskets, and water manifold. You should be able to notice a red or greenish color, or a residue that builds up at the leak. Belts and the fan should be checked for aging. Pressurize the system and watch hoses, radiator tank joints, and head gaskets for possible points of leakage. Test lubrication oil for water and obstructions at the radiator and fins. Remove and test the thermostat for proper opening and closing temperature, and replace as necessary. If the thermostats are installed in pairs and only one thermostat is found defective, both thermostats should be replaced. Verify that the correct temperature-range thermostat is being used.

Verify the operation of the low-coolant warning circuit and coolant thermistor.

Coolant freeze protection level can be checked with either a hydrometer or a refractometer. Some OEMs recommend use of a refractometer when checking freeze protection level. Most refractometers have both EG and PG scales, so make sure you are viewing the correct scale for the coolant being used. Coolants may be ethylene glycol based (EG), propylene glycol based (PG), or extended life coolants (ELCs). EG and PG coolants require close monitoring and the addition of supplemental coolant additives (SCAs). Monitoring is performed during a PM service of the engine. SCA test strips are dipped into the coolant and compared to the chart supplied. ELC coolants are low maintenance; coolant life is six years, with only one addition of SCA during that period. ELC coolants are only sold as a complete premixed coolant, which is dyed red. Do not dilute the coolant with additional water, and only add ELC premix as a make-up fluid. ELC coolants are incompatible with EG and PG coolants.

Fan operation will vary depending upon design. Some fans use a viscous clutch while others are hydraulically driven and speed is controlled by the engine ECM. Control of the fan may be through the use of a fanstat, by the engine management system, or by the vehicle air-conditioning. Some engine brakes also exercise control over the fan for additional braking. Compare fan operation with OEM service information.

Task A15   Check lubrication system for contamination, oil level, quality, temperature, pressure, filtration, and oil consumption; determine needed repairs.

Examination of lubrication oil can reveal problems associated with engine performance. An indication of oil dilution from coolant is water droplets forming on the dipstick while the engine is off, or if the oil appears gray in color after the engine has run. This can occur for a cracked block or cylinder head, blown head gasket, leaky oil cooler, or a leaky sleeve O-ring. Fuel in the oil will cause the oil to thin out and loose viscosity, potentially causing severe engine damage, and may appear on the dipstick as clearer oil at top of level indicator if the engine has not been run for a while. This problem may be caused by injector sleeve O-rings, dribbling fuel injectors, leaking fuel rails, or failed injection pump seals.

Excessive oil consumption can be an indicator of improperly seated or defective rings, a damaged piston, or defective valve stem seals. These problems will usually be accompanied by blue or black smoke in the exhaust. Low oil pressure may indicate a worn oil pump, restricted suction tube or screen, open, or improperly adjusted relief valve, clogged oil filter, or worn bearings. Regardless of the initial problem, if it is not corrected, more serious problems will appear.

Task A16   Check, record, and clear electronic diagnostic codes; monitor electronic data; determine needed repairs.

Current diesel engines use computerized engine control systems. According to SAE Standard J1930, diagnostic codes are known as Diagnostic Trouble Codes or DTCs. The DTC extraction method varies with the engine and vehicle manufacturer. You need to understand how to check, record, and clear DTCs. The typical process is to use a scan tool, diagnostic reader, or laptop computer to check for, record, and clear DTCs.

Modern engines are controlled and monitored by electronic modules. These modules are capable of storing fault codes. While each manufacturer may use its own code, it must design its software so that SAE standard codes are displayed. This means that one manufacturer's diagnostic hardware and software is capable of at least reading others. The term message identifier or MID is used to identify the chassis electronic system; the engine would be one of these systems, the transmission and antilock brakes others. Within the electronic system or MID would be varies branches, components, and subcircuits that are each divided into codes known as parameter identifiers (PIDs) and subsystem identifiers (SIDs). Within each PID or SID, faults are identified by common codes known as fault mode indicators or FMIs. This makes it easy for a technician trained on one equipment brand troubleshooting one of its competitors. For instance, FMI number 4 tells the technician that the component or circuit is voltage below normal or shorted low, and would be so on any PID or SID on any truck/bus manufactured in North America.

Electronic service tools (ESTs) connect to the truck/bus electronic systems by means of a data connection, either a 6-pin Deutsch connector (SAE 1587/1708) or a more recent 9-pin Deutsch connector known as an SAE J1939 connector. These connectors are usually referred to as ATA (American Trucking Association) connectors and are the same on all trucks/busses manufactured in North America. Most electronic systems will also blink codes out using dash-mounted diagnostic lights. More frequently today, digital driver dash interface is designed to produce readouts that keep the driver (and the technician) fully informed as to the status of all the monitored equipment on the truck/bus chassis.

Task A17   Perform visual inspection for physical damage and missing, modified or tampered components; determine needed repairs.

Experience teaches technicians how to visually inspect components with the kind of detail that enables the detection of tampering and modifications. A Detroit Diesel technician who at a glance can identify tampering on a Detroit Diesel engine may find it difficult to locate similar tampering abuse on a Cummins engine so there is no substitute for knowing the equipment on which you are working. Many cases of tampering and damage can be detected by looking for obvious signs such as missing components, frayed wiring or protective loom, electrical connectors not connected or hanging loose, and air, oil, or fuel leaks.

Technicians should be especially aware of tampering with any devices or components that have a bearing on engine emissions as this is illegal. Penalties for emission system infractions can be severe.

Task A18   Research applicable vehicle and service information, service precautions, and technical service bulletins; determine needed actions.

Technicians today use a variety of sources for service information to keep their skills current. Technological advancements are frequent, and so service procedures also need frequent revisions. The volume of information published each year is more than can be committed to memory. Where once a shelf of books was needed, now the same information is available on CD or DVD discs. A single compact disc is capable of containing more than 250,000 pages of text. New service information systems have been introduced to help technicians. Here is a partial list of sources for service information.

Maintenance manuals contain routine service intervals, procedures, and lube requirements for vehicle components and systems. Use these manuals for fluid recommendations, capacities, and charts. Torque values and adjustments can be found here, but not detailed repair information. You will need a service manual for that.

General and specialty repair manuals are manuals and databases produced by independent publishers. They compile information from various manufacturers into a single database for the service industry. Component manufacturers also publish service information in hard copy and electronic formats.

Service bulletins contain the latest information providing service tips, precautions, product improvements, recalls, field service modifications, and related service information for the benefit of the technician. Sometimes these will be updates to information previously published in service manuals, and take precedence over service manual information. Service bulletin manuals are usually available through the dealers, who receive them from the manufacturers. Check here for the most current information available.

Parts manuals not only provide information regarding which parts are correct, but also contain valuable illustrations that show the correct assembly of parts. Exploded views often show items that are not illustrated in the service manuals. Parts manuals are a useful guide for reassembling.

Internet-based information comes from many sources and contains a variety of documentation. Service information databases compiled by independent publishers are available by subscription over the Internet. Some manufacturers offer their dealers technician-training courses, service bulletins, recalls, and software upgrades that can be downloaded from the Internet. Most component suppliers to OEMs have web sites where information is also available. Other sources of information are the numerous interest groups and associations relating to almost all aspects of the transit industry. There are a number of very good technicians' associations for professional technicians. You can find them using your browser and favorite search engines.

B. Cylinder Head and Valve Train Diagnosis and Repair (5 Questions)

Task B1   Remove, inspect, disassemble, and clean cylinder head assembly(s).

Inspection of the cylinder head begins by checking the torque of cylinder head bolts before removing the head. This can indicate the potential for a warped or cracked head. Inspect the head gasket and head surface for signs of carbon or coolant stains between cylinders and passages before steam cleaning the head. Use a steam cleaner to remove oil, coolant, and carbon residue. Then, sand or peen the surface as necessary to remove all remaining carbon and gasket material. Use of sharp objects like knives or scrapers to remove residue may impair inspection for cracks and damage. When cracks are suspected but not apparent, magnetic or dye crack detection may be required to highlight cracks. Visually reinspect the head for cracks in the casting, scale build-up in the coolant passages, and pitting of machined surfaces. Excess accumulation of scale in water passages can usually be removed by placing the head in a soak tank for about two hours. Pressure testing the head can help to locate a crack in coolant and oil passages. Pressure testing cylinder heads is more effective when hot water is used.

Task B2   Inspect threaded holes, studs, and bolts for serviceability; service or replace as needed.

When removing or installing a component, carefully inspect all bolts, studs, holes, and nuts for damage to threads. Stripped, cross-threaded, nicked, rolled, and rusty threads can produce errors in the torque values. Use compressed air to remove foreign matter from holes. Obstructions lodged in holes, such as metal and liquids, can cause a bolt to bottom-out and reach torque value without obtaining the required clamping force. Hydrostatic pressure from liquids trapped in threaded holes can cause cracks when bolts are tightened. Examine the shank of bolts and studs for signs of twisting or overtorque fracturing. Some engine manufacturers require measuring the length of bolts before reuse. Replace defective bolts and studs. Ensure that nuts are not distorted or cracked, replace self-locking and defective nuts. Clean and dress threads in holes. If threads are damaged beyond use, drill and tap the hole and install an approved helicoil. Use an approved stud extractor or Eze-out to remove broken studs and bolts.

Steps in the installation of a helical screw repair coil: (A) Drill the damaged threads using the correct size drill bit. Clean all metal chips out of the hole. (B) Tap new threads in the hole, using the specified tap. The thread depth should exceed the length of the bolt. (C) Install the proper-size coil insert on the mandrel provided in the installation kit. Bottom it against the tang. (D) Lubricate the insert with oil and thread it into the hole until flush with the surface. Use a punch or side cutter to break off the tang.

Task B3   **Measure cylinder head-to-deck thickness, and check mating surfaces for warpage and surface finish; inspect for cracks and damage; check condition of passages; inspect core and gallery plugs; service as needed.**

The cylinder head should be checked for warping. Using a straightedge and feeler gauge, measure the face (fire deck) for flatness. Check for transverse (across the width of a cylinder head) warpage at each end and between each cylinder. Check for longitudinal (across the length of the cylinder head) warpage at points above and below each cylinder, between center centerlines of valves, and the outer edge of the fire deck. Compare measurements to manufacturer's specifications to determine the need for resurfacing. Resurfacing may necessitate removing inserts, such as valve guides, and/or deburring of coolant ports. Check cylinder head thickness, called deck-to-deck thickness, to determine the maximum allowable machining limit.

Task B4   **Inspect valves, guides, seats, springs, retainers, rotators, locks, and seals; determine serviceability and needed repairs.**

Check valve spring straightness with a T-square. This will help prevent stems from binding and twisting in guides as valves are operated. Uneven side pressure causes valve stems and guides to wear faster. Also check the unloaded, or free length, and loaded length, or tension, of the spring using a spring tension gauge. Replace valve springs that do not meet specifications to prevent damage to the engine. Valve springs using the same valve bridge must be replaced together to prevent unbalanced valve operation. This can damage the bridge from uneven stress, cause valve guides and stems to wear faster, and result in insufficient scavenging of the cylinder.

Valve spring retainers keep the valve spring on the valve. Valve rotators are used to rotate the valve each time it is actuated. Some engines also have seals. Careful inspection of these components can prevent damage and extend the life of the engine. Retainers, also called keepers, that become worn or damaged may allow the spring to release the valve. If this happens, the valve drops into the cylinder and jams the piston, a condition called dropping a valve. Free-release rotators use engine vibration and exhaust gases to rotate, because the spring tension is removed when it is depressed. Mechanical rotators apply a torque to rotate the valve each time the valve is opened. Both will help eliminate carbon deposits from forming on the valve face and seats. Seals between the valve guide and valve stem prevent oil from entering the cylinder on the intake stroke. Defective seals can cause oil consumption, leading to excess smoking, carbon build-up, fouling of injectors, formation of hot spots, and precombustion or dieseling.

After a cylinder head has been checked or resurfaced and is considered useable, the valve guides should be checked for wear. Check the guide inside diameter with a snap, ball, or dial gauge in three different locations throughout the length of the guide. If guides are worn excessively, they should be replaced, if they are the replaceable types. In most cylinder heads, the valve guides can be removed by using a driver. Select a driver that fits the valve guide. After selection of a driver the guide should be removed as follows:

- Support the cylinder head on a block or stand.

- If guide is to be pressed out, place the cylinder head in the press and align the driver with the press ram.

- Note the position of the guide in relation to the cylinder head for reference when installing a new guide.

- Drive or press the guide out of cylinder head making sure that the driver is driven or pressed straight. If it is not pressed straight, damage to the valve guide or cylinder head may result.

- After the guide has been removed, check the guide bore for scoring. A badly scored guide bore may have to be reamed out to accommodate the next larger size guide.

- After the guide bore is checked, select the correct guide (intake or exhaust) and insert it in the guide bore.

- Insert the guide driver and drive or press the guide into the cylinder head until the correct guide position is reached. If manufacturer's specifications are not available, drive the guide into the same position as the old guide.

- Many engine manufacturers recommend that a new guide be reamed after installation to ensure that the guide inside diameter did not change during the installation process. Never ream a surface-hardened guide.

Valve seats are used to absorb the forces of valves opening and closing, and the effects of expansion and contraction. Properly inspect, clean, and recondition or replace valve seats to maintain a good valve-to-head seal. Loose valve seats are located by lightly tapping each seat with the peen end of a ball-peen hammer. A loose seat will produce a sound that differs from the ring sound created by tapping the head adjacent to the seat. The seat may even move a little. Leaks may occur if the seat has visible cracks, uneven burns or discoloration, or an opening, which is oblong or excessive in diameter. Replace any seats that cannot be reconditioned. To recondition a valve seat, a mandrel pilot is installed in the valve guide to keep a specially dressed grinding wheel centered in the seat. The grinding wheel may be dressed with a 1° steeper angle than the valve. This is known as an interference angle and helps the valve to cut through future carbon build-up. An interference angle is never used on engines equipped with valve rotators. New seats must be dressed for the valve that will use them. Repeat the inspection process after reconditioning the valve seats. Also, check the concentricity (roundness) of the seat opening. If the seat is not round, it will not seal during compression and combustion strokes. Verify that the valve head height is correct. If the head height is out of specification, the seats may need to be replaced, as this may alter the compression ratios or injector spray pattern. Do not remove any more material than is necessary to refinish the valve seat.

After guides have been replaced or reconditioned, valve seat checking and reconditioning should be done. If the seat passes all checks, it should be reconditioned as outlined, using a specially designed valve seat grinder. Most valve seat grinders are similar to the one in the figure. Select the proper mandrel pilot by measuring the valve stem with a micrometer or caliper or by referring to manufacturer's specifications.

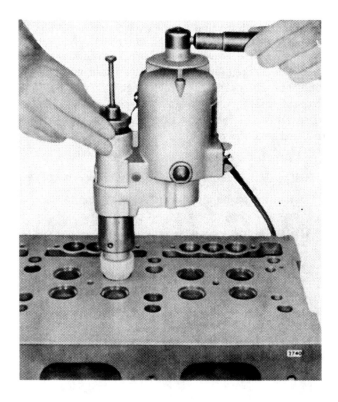

Before installing valves, check the valve-to-seat contact. One method of doing this is to coat the face of the valve with Prussian blue and lightly snap the valve into its seat. Remove the valve and check its face. If the valve face does not have an even seating profile, the seat is not concentric and must be reground. When the valves have been seated, check the valve heads height. This checks the distance the valve protrudes above or below the machined surface of the head. Place a straightedge across the cylinder head and use a feeler gauge to measure distance between the valve head and straightedge. If the valve is designed to protrude above the head, place the straightedge on the valve. Use a feeler gauge between the valve head and the cylinder head machined surface. A depth micrometer may also be used for this procedure.

Task B5   Inspect, reinstall or replace injector sleeves and seals; pressure test to verify repair (if applicable); measure injector tip or nozzle protrusion where specified by manufacturer.

If a leak is detected in an injector sleeve seal, the seal must be replaced to prevent further damage to the engine. After following the manufacturer's procedures for removing and reinstalling the injector sleeve, perform a cylinder head pressure test to ensure the sleeve is sealed. Check the injector tip height to ensure it meets manufacturer specifications. Incorrect injector height can be corrected by either shimming the injector, or reaming the injector sleeve. Excessive protrusion on some engines can lead to injector tip-to-piston interference, causing engine/injector damage. Failure to do so may also result in incomplete combustion due to an improper injector spray pattern.

Task B6   Inspect, reinstall or replace valve bridges (crossheads) and guides; adjust bridges (crossheads).

Valve crossheads are used on diesel engines that use three and four valves per cylinder. Four valve heads used on 4-stroke cycle engines have two intake and two exhaust valves per cylinder. Two-stroke cycle engines, such as Detroit Diesel, use four exhaust valves in each cylinder. The crosshead is a bracket or bridge-like device that allows a single rocker arm to open two valves at the same time. The crossheads (bridge) must be checked for wear as follows:

- Visually check crosshead for cracks and with magnetic crack detector if available.
- Check crosshead inside diameter for out-of-roundness and excessive diameter.
- Visually check for wear at the point of contact between the rocker lever and the crosshead.
- Check the adjusting screw threads for damaged or worn threads.
- Check the crosshead guide pin diameter with a micrometer.
- Check the crosshead guide pin to ensure that it is at right angles to the head-milled surface.

Task B7   Clean components; reassemble, check, and install cylinder head assembly as specified by the manufacturer.

The assembly of the cylinder head or heads onto the engine block will involve many different procedures that are distinct to a given engine. The following procedures are general in nature and are offered to supplement the manufacturer's service manual.

Before attempting to install the cylinder head, make sure the cylinder head and block surface are free from all rust, dirt, old gasket materials, and grease or oil. Before placing the head gasket on the block, make sure all bolt holes in the block are free of oil and dirt by blowing them out with compressed air.

Select the correct head gasket and place it on the cylinder block. Check the head gasket closely as some have an "up" or "top" mark. In most cases head gaskets will be installed dry with no sealer, although in some situations an engine manufacturer may recommend applying sealer to the gasket before cylinder head installation. Check the service manual for installation recommendations.

After setting the head gasket on the block, place coolant and oil O-rings (if used) in the correct positions. Some head gaskets will be of a one-piece solid composition type, while others will be made of steel and composed of several sections or pieces. If the fire ring is not integral with the head gasket, ensure that it is correctly positioned on the liner flange. Installation of some head gaskets requires the use of a dowel (threaded rods) that are screwed in the head bolt holes to hold the head gasket in place during head assembly. If dowels are recommended, they can be made from bolts by sawing off the heads and machining a taper on the end. Most modern engines will have dowels or locating pins in the block to aid in holding the cylinder head gasket in place during cylinder head assembly.

After the gasket and all O-rings are in place, check the cylinders to make sure no foreign objects have been left in them, such as O-rings and bolts.

Place the head(s) on the cylinder block carefully to avoid damage to the head gasket. On in-line engines with two or three separate heads that do not use dowel pins, it may be necessary to line up heads by placing a straightedge across the intake or exhaust manifold surfaces.

Clean and inspect all head bolts or cap screws for erosion or pitting. Coat bolt threads with engine oil and place the bolts or cap screws into the bolt holes in the head and the block.

Using a speed wrench and the appropriate-size socket, start at the center of the cylinder head and turn the bolts down snug.

When all bolts have been tightened this far, continue to progressive torque each bolt in sequence with a torque wrench until the recommended final torque is achieved. Depending on manufacturer's procedures, retorquing of the head bolts may be required. Once the head bolts are properly torqued, final installation of the valvetrain mechanisms, and adjustments of it will follow. Once repairs are complete, the engine should be run up to normal operating temperature, and checked for leaks.

Task B8   Inspect, measure, reinstall, or replace pushrods, rocker arms, rocker arm shafts, and supports for wear, bending, cracks, looseness, and blocked oil passages. Visually inspect for wear and correct routing.

After the cylinder heads have been tightened to the correct specification, the pushrods or tubes and rocker arm assemblies may be installed. Check each push tube for straightness. Straightening of push tubes is discouraged. Bent ones should be replaced. Push tubes must be checked for breaks or cracks where the ball socket on either end had been fitted into the tube. Place rod or tubes in the engine, making sure that they fit into the cam followers or tappets. The rocker arm assembly or rocker arm is one component that is occasionally overlooked during diesel engine overhaul. Check each boss for wear. Oil consumption occurs because the increased clearance allows excessive oil to splash or leak on the valve stem. This oil will run down the valve and end up in the combustion chamber.

Task B9   Inspect, install, and adjust cam followers.

Tappets, lifters, and cam followers are components that are positioned to ride, or be actuated by, a cam profile. The term tappet has a broader definition and is sometimes used to describe what is more often referred to as a rocker lever. Because North American engine and fuel system original equipment manufacturers (OEMs) use all three terms, the technician should be familiar with them. In this text, the term tappet will not be generally used to describe a rocker and the term follower will be generally used to describe a component that rides the cam profile, except when describing specific OEM systems where an alternative term is preferred. The function of cam followers is to reduce friction and evenly distribute the force imparted from the cam's profile to the train for which it is responsible for actuating. Diesel engines using cylinder block-mounted camshafts use two categories of follower, while those using overhead camshafts use either direct actuated rockers or roller-type cam followers.

Task B10   Adjust valve clearances and injector settings.

When valves are properly adjusted, there should be clearance between the pallet of the rocker arm and the top of the valve stem or valve bridge. Valve lash is required because as the moving parts gain heat they expand. If clearance were not factored somewhere in the valvetrain, the valves would remain open by the time an engine reached operating temperature. Actual valve lash values depend on factors such as the length of the push tubes and the materials used in valve manufacture. Exhaust valves are subject to more heat and as a consequence, OEMs require a valve lash setting for exhaust valves that is usually greater than the intake valve lash setting.

Maladjusted Valves

A loose valve adjustment will retard valve opening and advance valve closing, decreasing the cylinder breathing time. Actuating cam geometry is designed to provide some "forgiveness" to the train at valve opening and valve closure to reduce the shock loading to which the train is subject. When valves are set loose, the valvetrain is loaded at a point on the cam ramp beyond the intended point. The same occurs at valve closure when the valve is seated. High valve opening and closing velocities subject the valve and its seat to hammering, which can result in cracking, failure at the head to stem fillet, and scuffing to the cam and its follower.

Valve Adjustment Procedure

The following steps outline the valve adjustment procedure on a typical 4-stroke cycle, in-line 6-cylinder diesel engine. Valves should always be adjusted using the OEMs specifications and procedure.

Caution: Shortcutting the engine OEM recommended valve (and injector) setting can result in engine damage unless the technician knows the engine well. Cam profiles are not always symmetrical and some engines may have cam profiles designed with ramps between base circle and outer base circle for purposes such as actuating engine compression brakes. Similarly, a valve rocker that shows what appears to be excessive lash when not in its setting position is not necessarily defective.

Six-cylinder engine firing order: 1-5-3-6-2-4 cylinder throw pairings: 1-6 @ TDC 5-2 @ 120 degrees BTDC 3-4 @ 120° ATDC

Cylinder throw pairings are sometimes called companion cylinders. In other words, when #1 piston is at TDC completing its compression stroke, #6 piston (its companion) is also at TDC having just completed its exhaust stroke. If the engine is viewed from overhead with the rocker covers removed, engine position can be identified by observing the valves over a pair of companion cylinders. For instance, when the engine calibration plate indicates that the pistons in cylinders #1 and #6 are approaching TDC and the valves over #6 are both closed (lash is evident), then the point at which the valves over #1 cylinder rock (exhaust closing, intake opening) at valve overlap will indicate that #1 is at TDC having completed its exhaust stroke and #6 at TDC having completed its compression stroke. This method of orienting engine location is commonly used for valve adjustment.

Adjustment

1. Locate the valve lash dimensions. The lash specification for the exhaust valve(s) will usually (but not always) be greater than that for the inlet valve.

2. Valves on current diesel engines should usually be set under static conditions and with the engine coolant 100°F (37°C) or less. Locate the engine timing indicator and the cylinder calibration indexes, 120° apart: depending on the engine, this may be located on a harmonic balancer, any pulley driven at engine speed, or the flywheel.

3. Ensure that the engine is prevented from starting by mechanically or electrically no-fueling the engine. The engine will have to be barred in its normal direction of rotation through two revolutions during the valve setting procedure, requiring the engine to be no-fueled to avoid an unwanted startup.

4. If the engine is equipped with valve bridges or yokes that require adjustment, performed this procedure prior to the valve adjustment. To adjust a valve yoke, back off the rocker arm then loosen the yoke adjusting screw locknut and back off the yoke adjusting screw. Using finger pressure on the rocker arm (or yoke), load the pallet end (opposite to the adjusting screw) of the yoke to contact the valve. Next, screw the yoke adjusting screw CW until it bottoms on its valve stem. Turn an additional one flat of a nut (1/6 of a turn), then lock to position with the locknut.

Caution: When loosening and tightening the valve yoke adjusting screw locknut, the guide on the cylinder head is vulnerable to bending. Most OEMs recommend that the yoke be removed from the guide and placed in a vice to back off and final torque the adjusting screw locknut.

5. To verify that the yoke is properly adjusted, insert two similarly sized thickness gauges of 0.010 in. or less between each valve stem and the yoke. Load the yoke with finger pressure on the rocker arm and simultaneously withdraw both thickness gauges. They should produce equal drag as they are withdrawn. If the yokes are to be adjusted, they can be adjusted in sequence as each valve is adjusted. In some engines, valve yokes can only be adjusted with the rocker assemblies removed as it is impossible to access the yoke to verify the adjustment otherwise.

6. If the instructions in the OEM literature indicate that valves must be adjusted in a specific engine location, ensure that this is observed: The cams that actuate the valves may only have a small percentage of base circle. Clearances should be checked prior to making any adjustments. If the clearances are within specifications, no adjustment is required. Setting valves requires the lash dimension between the rocker arm and the valve stem on the rocker arm and the valve yoke to be defined. When performing this procedure, the valve adjusting screw locknut should be backed off and the adjusting screw backed out. Insert the specified size of thickness gauge between the rocker and the valve stem/yoke. Release the thickness gauge. Then turn the adjusting screw CW until it bottoms. Turn an additional 1/2 flat of a nut (1/12 turn). Hold the adjusting screw with a screwdriver and with a wrench, lock into position. Now for the first time since inserting the thickness gauge, handle it once again by withdrawing the thickness gauge. A light drag indicates that the valve is properly set. If the valve lash setting is either too loose or too tight, repeat the setting procedure. Do not set valves too tight. Set all the valves in cylinder firing order sequence rotating the engine 120° between setting. It is preferable to begin at #1 cylinder and proceed through the engine in firing order sequence.

Task B11   Inspect, measure, and reinstall or replace overhead camshaft and bearings; measure and adjust endplay and backlash.

A careful inspection of the camshaft will ensure that proper valve and injector timing can be obtained. Because the camshaft is constantly supplied with oil, cleaning should be as simple as rinsing with dip tank solution and blowing out with compressed air. Oil passages may require cleaning with a wire brush. Discard any camshaft that does not pass a visual inspection. If the camshaft passes visual inspection, take measurements and compare with manufacturer's specifications. The cam lobes must be inspected for signs of pitting, scoring, or flat spots. Use the manufacturer's procedures for taking lobe measurements. The heel-to-toe measurement is equal to the diameter of the circular portion of the lobe and the maximum amount of lift created by the rise at the top of the lobe. Typically, the cam follower will ride in the center of lobe travel causing a groove to form. This groove should not be greater than OEM specifications. If either the top of the lobe or the circular portion is worn (grooved), the amount of lift will be reduced. If an intake or exhaust cam lobe fails to produce enough lift, the valve it controls may not open enough to supply adequate air for combustion to occur. If the lobe actuates an injector, the cylinder may not fire. Visually inspect bearing journals for bluing, scoring, or wear. Measure the diameter of journals with an outside micrometer, and look for out-of-round condition by checking around the journal in several places. Inspect the shaft and gear keyway for cracks or distortion. Ensure the keyway in the gear and shaft matches the woodruff key. Any movement between these three can randomly change the timing of valves and injectors. A shift of as little as 0.004 in. (0.1 mm) in the keyway could equate to almost 1° in timing shift.

Some manufacturers use an offset key when installing the cam gear to the camshaft to obtain desired timing. This should not be confused with a key that is partially sheared. When an offset timing key is used, make sure that the original key or an identical replacement is installed with the offset facing in the correct direction.

C. Engine Block Diagnosis, Repair and Overhaul (5 Questions)

Task C1   **Remove, inspect, service, and reinstall pans, covers, breathers, gaskets, seals, and wear rings.**

Oil pans are usually located in the airflow under the frame rails and are vulnerable to damage from objects on the road from rocks on rough terrain to small animals on the highway. Most highway diesel engines have oil pans that can be removed from the engine while it is in chassis. Drain the oil sump prior to removing it. Oil pans that seal to the engine block using gaskets can be difficult to remove especially where adhesives have been used to ensure a seal. A 4-lb. rubber mallet may facilitate removal. Avoid driving screwdrivers between the oil pan and its block mating flange because this can result in damage to the oil pan flange and the cylinder block mating flange. Oil pans that use rubber isolator seals tend not to present removal problems. After removal the oil pan should be cleaned of gasket residues and washed with a pressure washer. Inspect the oil pan mating flanges, check for cracks, and test the drain plug threads. Cast aluminum oil pans that fasten both to the cylinder block and to the flywheel housing are prone to stress cracking in the rear. With these oil pans always meticulously observe torque sequences and values. Cast aluminum oil pans can be successfully repair welded without distortion using the TIG (tungsten inert gas) process, providing the cracks are not too large and the oil-saturated area of the crack is ground clean. It is more difficult to execute a lasting repair weld on a stamped steel oil pan due to the distortion that results. Steel oil pans may rust through so they should be replaced rather than welded. Whenever an oil pan has indications of porosity resulting from corrosion, it should always be replaced. The technician should always be aware of the fact that a failure of the oil sump or its drain plug on the road usually results in the loss of the engine.

Task C2   **Disassemble, clean, and inspect engine block for cracks; check mating surfaces for damage or warpage and surface finish; check deck height; check condition of passages, core, and gallery plugs; inspect threaded holes, studs, dowel pins and bolts for serviceability; service, reinstall or replace as needed.**

The block counterbore area must be completely cleaned of all rust, scale, and grease. Most scale and/or rust can be removed using a piece of crocus cloth or 100- to 120-grit emery paper or wet-dry sandpaper. Inspect the counterbore closely for cracks, and ensure that no rough or eroded areas exist to ruin a sleeve or cut its O-ring. Damage of this nature may require resleeving the counterbore by an experienced technician with special tools. If the counterbore is in good shape, use a dial indicator mounted on a fixture to measure the counterbore depth. A depth micrometer can be used in place of the dial indicator if firm contact is maintained with the block surface.

Check for deck warpage using a straightedge and thickness gauge. A typical maximum specification is in the region of 0.1 mm or 0.004 in.

Check the main bearing bore and alignment. Check the engine service history to ensure that the engine has not previously line bored. A master bar is used to check alignment for a specific engine series. Check that the correct bar has been selected for the engine being tested and then clamp to position by torquing down the main caps minus the main bearings. The master bar should rotate in the cylinder block main bearing line bore without binding. If it binds, the cylinder block should be line bored.

Check the cam bore dimensions and install the cam bushings with the correct cam bushing installation equipment. Use drivers with great care as the bushings may easily be damaged and ensure that the oil holes are lined up prior to driving home the bushing.

Gallery and expansion plugs are often interference fitted and sealed with silicone, thread sealants, and hydraulic dope. Electromagnetic flux test the block for cracks at each chassis overhaul.

Task C3   **Inspect cylinder sleeve counterbore and lower bore; check bore distortion; determine needed service.**

Check the cylinder sleeve counterbore for the correct depth and circumference. Counterbore depth should typically not vary by more than 0.025 mm (0.001 in.). Counterbore depth can be subtracted from the sleeve flange dimension to calculate sleeve protrusion. Recut and shim the counterbore using OEM-recommended tools and specifications.

Task C4   **Inspect and measure cylinder walls or liners for wear and damage; determine needed service.**

Before reinstalling, used wet sleeves should be cleaned by immersion or glass-bead blasting and closely inspected. Sleeves that fail the inspection should be discarded to ensure maximum overhaul life. The inspection should include checking for taper using an inside micrometer, snap gauge, or dial bore gauge. An out-of-round condition can be detected by measuring the diameter at one point inside the cylinder and rotating the inside micrometer around the cylinder 180°. To check liner to block height, first install and leave a liner clamping tool in place, holding the liner down. Next, measure the liner protrusion with a sled gauge. The liner protrusion should be 0.000 to 0.003 in. with no more than a 0.002-in. variance between liners under one cylinder head.

After cleaning and pressure testing, inspect the cylinder block. Since most engine cooling is accomplished by heat transfer through the cylinder liners to the water jacket, good liner-to-block contact must exist when the engine is operating. Whenever the cylinder liners are removed from an engine, the block bores must be inspected.

Two basic methods of liner reconditioning are used: glaze busting and honing.

Glaze busting: When checked to be within serviceability specs, the liner should be deglazed. Deglazing involves the least amount of material removal. A power-driven flex hone or rigid hone with 200–250-grit stones is used. The best type of glaze buster is the flex hone, typically a conical (Xmas tree) or cylindrically shaped shaft with flexible branches with carbon/abrasive balls. The objective of glaze busting is to machine away the cylinder ridge above the ring belt travel and reestablish the crosshatch. The drill should be set at 120–180 rpm and used in rhythmic reciprocating thrusts. Short sequences, stopping frequently to inspect the finish produce the best results. You should observe a 60° crossover angle, 15–20 microinch crosshatch.

Honing: Use a hone in which the cutting radius of the stones can be set in a fixed position to remove irregularities in the bore rather than following the irregularities as with a spring-loaded hone. Clean the stones frequently with a wire brush to prevent stone loading. Follow the hone manufacturer instructions regarding the use of oil or kerosene on the stones. Do not use such cutting agents with a dry hone.

Insert the hone in the bore and adjust the stones snugly to the narrowest section. When correctly adjusted, the hone will not shake in the bore, but will drag freely up and down the bore when the hone is not running.

Start the hone and "feel out" the bore for high spots that will cause an increased drag on the stones. Move the hone up and down the bore with short overlapping strokes about 1-in. long. Concentrate on the high spots in the first cut. As these are removed, the drag on the hone will become lighter and smoother. Feed the hone gently to avoid an excessive increase in the bore diameter. Some stones cut rapidly even under low tension.

When the bore is clean, remove the hone, inspect the stones, and measure the bore. Determine which spots must be honed most. Moving the hone from the top to the bottom of the bore will not correct an out-of-round condition. To remain in one spot too long will cause the bore to become irregular. Where and how much to hone can be judged by feel. A heavy cut in a distorted bore produces a steady drag on the hone and makes it difficult to feel the high spots. Therefore, use a light cut with frequent stone adjustments. Overhead boring tools can be set to produce the required stroke rate for the specified crosshatch but if using a hand-held tool, remember that a few short strokes with a moderate radial load tend to produce a better crosshatch pattern than many strokes with a light radial load. You should observe a 60° crossover angle pattern, 15–20 microinch crosshatch. It should be clearly visible by eye. Wash the cylinder block thoroughly with soap and water after the honing operation is completed.

Note: Some manufacturers do not recommend resurfacing a liner, they recommend replacement only.

Task C5   **Replace/reinstall cylinder liners and seals; check and adjust liner heights.**

Selective fitting of liners to block bores is good practice whether or not the block bore has been machined. To selective fit a set of dry liners to a block, measure the inside diameter of each block bore across the north-south and east-west faces and grade in order of size. Next, get the set of new liners and measure the outside diameter of each; once again grade in order of size. Ensure that every measurement falls within OEM specifications. Then fit the liner with the largest OD to the block bore with the largest ID. Loose fits of dry liners around 0.035 mm (0.0015 in.) are common.

To pull wet-type sleeves from the cylinder block:

- Select the correct adapter plate that will fit the sleeve.

- Make sure the plate fits snugly in the sleeve (to prevent cocking) and that the outside diameter of the puller plate is not larger than the sleeve outside diameter. An adapter plate larger than the sleeve may damage the block.

- Attach the adapter plate to the through bolt.

- If the adapter plate is the type that can be installed from the top of the sleeve, it will have a cutaway or milled area on each side. This, along with the swivel on the bottom of the through bolt, allows the plate to be tipped slightly and inserted from the top, eliminating the need to install the adapter plate in the bottom of the sleeve and then inserting the through bolt and attaching a nut.

- After you install the adapter plate, hold the through bolt and adapter plate firmly in the sleeve with one hand and install the support bracket with the other hand.

- Screw the nut down on the through bolt until it contacts the support bracket. This will hold the adapter plate and through bolt snugly in place.

- Before tightening the sleeve puller nut, make sure the sleeve puller supports or legs are positioned on a solid part of the block. Tighten the nut with a ratchet; the sleeve should start to move upward. If it does not and the puller nut becomes hard to turn, stop and recheck your puller installation before proceeding.

- On wet-type sleeves, after the sleeve has been pulled from the block far enough to clear the O-rings, tip or swivel the sleeve puller adapter plate and remove the sleeve puller assembly.

- The sleeve can now be lifted out by hand.

- Engines with tight fitting dry-type sleeves may require a special hydraulic puller.

Once the liners have been removed, the block should be carefully inspected. If the engine had wet liners, inspect the liner seal area of the block for damage. The O-ring band area is subject to grinding action when debris from liner cavitation is caught between the liner and the block. The counterbore should be cleaned and checked for cracks, first by a visual inspection and then by magnetic particle testing. If the cylinder block counterbore is damaged, it can be machined and fitted with an oversize liner, or by machining and installing a sleeve. Measure and record each counterbore and each liner flange thickness. Sort the liners so that those with the largest dimension flange may be installed in the deepest counterbore so that manufacturing tolerances are minimized. Temporarily install the liners without seals, install clamping fixtures and measure liner protrusion. It may be necessary to switch liner locations or install shims until liner protrusion is within manufacturer's specifications. When the block and the liners are ready for installation, install the liner seals and apply the lubricant recommended by the manufacturer. Install the liners, recheck liner protrusion, and check the clearance between the liner and its lower bore to ensure that no distortion has occurred during installation.

Task C6   **Inspect in-block camshaft bearings for wear and damage; replace as needed.**

The camshaft is supported by pressure-lubricated, friction bearings at main journals. The camshaft is subjected to loading whenever a cam is actuating the train that rides its profile. This loading can be considerable especially when the camshaft actuates engine compression brake hydraulics and fuel injection pumping apparatus. Cam bushings are normally replaced during engine overhaul, but if they are to be reused, they must be measured with a dial bore gauge or telescoping gauge and micrometer to ensure that they are within the OEM reuse parameters. Interference fit, cylinder block located, camshaft bushings are removed sequentially starting usually from the front to the back of the cylinder block using a correctly sized bushing driver (mandrel) and slide hammer. Interference fit cam bushings are installed using the same tools used to remove them. Care should be taken to properly align the oil holes. When installing bushings to a cylinder block in chassis where access and visibility are restricted, painting the oil hole location on the bushing rim with White-Out (correction fluid) may help align the bushing before driving it. Ensure that the correct bushing is driven into each bore. Bushings and their support bores vary dimensionally and new bushings rarely survive being driven into a bore and then removed. Bearing clearance should be measured after installation. Bearing clearance is the camshaft bushing dimension measured with a dial bore gauge or telescoping gauge and micrometer subtracted from the camshaft journal dimension measured with a micrometer.

Task C7   **Inspect, measure, reinstall or replace in-block camshaft; measure and adjust endplay; inspect, reinstall or replace, and adjust cam followers (if applicable).**

A careful inspection of the camshaft will ensure that the specified valve and injector timing can be obtained. Because the camshaft is constantly supplied with oil, cleaning should be as simple as rinsing with dip tank solution and blowing out with compressed air. Oil passages may require cleaning with a wire brush. Discard any camshaft that does not pass a visual inspection. If the camshaft passes visual inspection, take measurements and compare with manufacturer's specifications. Cam lobes must be inspected for signs of pitting, scoring, or flat spots. Follow the manufacturer's procedures for taking lobe measurements. The heel-to-toe measurement is equal to the diameter of the circular portion of the lobe and the maximum amount of lift created by the rise at the top of the lobe. Typically, the cam follower will ride in the center of lobe travel causing a groove to form. This wear groove should not exceed the manufacturer's specifications. If either the top of the lobe or the circular portion is worn (grooved), lift will be reduced. If an intake or exhaust cam lobe fails to produce enough lift, the valve it controls may not open enough to supply adequate air for combustion to occur. If the lobe actuates an injector, the injector may not deliver the proper amount of fuel. Visually inspect bearing journals for bluing, scoring, or wear. Measure the diameter of journals with an outside micrometer, and look for out-of-round condition by checking around the journal in several places. Inspect the shaft and gear keyway for cracks or distortion. Ensure the keyway in the gear and shaft matches the woodruff key. Any movement between these three can randomly change the timing of valves and injectors. A shift of as little as 0.004 in. (0.1 mm) in the keyway could equate to almost one degree in timing shift.

Camshaft endplay is defined either by free or captured thrust washers/plates. Thrust loads are not normally excessive unless the camshaft is driven by a helical-toothed gear, in which there is more likely to be wear at the thrust faces. Endplay is best measured with a dial indicator. The camshaft should be gently levered longitudinally rearward then foreword and the travel measured and checked to specifications.

Task C8   **Clean and inspect crankshaft and journals for surface cracks and damage; check condition of oil passages; check passage plugs; measure journal diameters; check mounting surfaces; determine needed service.**

An inspection of the crankshaft begins with the connecting rod and main bearing shells. The bearing shells will usually show signs of damage before the crank. If the bearings indicate damage, inspect the crankshaft journal associated with that shell before it is cleaned. Clean the crank and visually inspect for obvious damage. This includes cracked or worn front hub key slots, the main bearing journal for scoring or bluing, dowel pins and holes for cracks and wearing, oil supply holes for cracks, and the shaft for seal grooving. In addition, a careful inspection of the crank is performed by taking measurements. An out-of-round journal can be detected by using an outside micrometer to measure in at least two places around the journal's diameter. Generally, this measurement should not exceed 0.002 in. (0.051 mm). To identify a tapered journal, measure the diameter nearest each side and the middle of the journal. Examples of these measurements are no more than 0.0015 in. (0.381 mm) between any two, no more than 0.0005 in. (0.013 mm) between tapers in the journal and bearing. Check thrust surfaces visually for scoring, nicks, and gouges. Measure clearance with an inside micrometer. Visually inspect for obvious cracks or breaks, and use either magnetic flux detection to find smaller, less obvious ones.

Crankshaft Failures

Bending failures: Abnormal bending stresses occur when:

- Main bearing bores are misaligned. Any kind of cylinder block irregularity will cause abnormal stress over the affected crankshaft area.

- Main bearings fail or become irregularly worn.

- Main caps are broken or loose.

- Standard specification main bearing shells are installed where an oversize is required.

- The flywheel housing is eccentrically (relative to the crankshaft) positioned on the cylinder block. A possible outcome of failure to indicate (using a dial indicator) a flywheel housing on reinstallation.

- Crankshaft is not properly supported either out of the engine prior to installation or while replacing main bearings in-chassis; the latter practice is more likely to damage the block line bore but may deform the crankshaft as well.

- Bending failures tend to initiate at the main journal fillet and extend through to throw journal fillet at 90° to the crankshaft axis.

Torsional Failures

Excessive torsional stress can result in fractures occurring from an initiation point in a journal oil hole extending through the fillet at a 45° angle or circumferential severing through a fillet. In an in-line 6-cylinder engine, the #5 and 6 journal oil holes tend to be vulnerable to torsional failure when the crankshaft is subjected to high torsional loads.

Causes of crankshaft torsional failures:

- Loose, damaged, or defective harmonic damper or flywheel assembly.

- Unbalanced engine driven components, such as fan pulleys and couplings, fan assembly, idler components, compressors, and PTOs.

- Engine overspeed. Even fractional engine overspeeding will subject the crankshaft to torsional stresses that it was not engineered for that may initiate a failure. When performing failure analysis, remember that the event that caused the failure and the actual failure may be separated by a considerable time span.

- Unbalanced cylinder loading. A dead cylinder or fuel injection malfunction of overfueling or underfueling a cylinder(s) can result in a torsional failure.

- Defective engine mounts. This can produce a "shock load" effect on the power train.

Task C9   **Inspect, reinstall or replace main bearings; check cap fit and bearing clearances; check and correct crankshaft endplay.**

Inspection of the main bearing bore alignment and out-of-roundness can prevent possible damage to the crankshaft or premature bearing failure. If the bearing bore is out of alignment, uneven pressure is applied to each of the bearings, causing the crank to flex as it is rotated. This flexing and its effects can also be created by an operator who "hot rods" or when the engine becomes lugged down. In this case, the engine will heat unevenly, resulting in uneven expansion of the journal. Verifying the alignment using a master bar may prove the latter. This flexing can cause the crank to fracture and will result in bearings with unusual wear patterns. Use an inside micrometer to check the bearing bore for correct diameter and an out-of-round condition. If the bearing bore is too large, bearing shells will not seat properly against the bore. This may cause inadequate heat transfer from the shells to the bore, turning, or movement of shells in the bore, and misalignment of the crank when load is applied. An out-of-round bore will cause uneven torque on the crank, which could lead to fatigue, fracturing, and failure or breakage of the crank. The bearing shells will not be lubricated evenly and will wear more on one side, while the distorted shell will create fractures in the bearing surface.

Correct installation of bearing shells is essential to any successful overhaul. About 13 percent of premature bearing failures are caused by improper assembly of bearings. After reconditioning the block and selecting proper-sized bearings, carefully install the top half of each bearing into the cylinder block. Ensure the locating lugs (indented notch in each shell) are fitted into the matching slot in the cylinder block bore. This type of error can cause movement of bearings leading to metal in oil and finally a spun bearing causing oil starvation and seizing of the crank. Visually inspect the oil passage in the shell to ensure that it is aligned to the oil passage in the bore. Failure to align the oil passages will result in oil starvation, bearing fatigue, scoring, metal in oil, and finally, other bearing failure. Carefully install the rear main seal (split type), crank in the block, and align marks on the crank timing gear and cam gears. Use a Plastigage® to check all main bearing clearances. If clearances are not within specification, check for improperly sized bearings, dirt, or metal under shells, or misalignment of shells. If clearances are correct, remove the Plastigage, lubricate bearings, reinstall bearing caps, and torque to specifications. Rotate the crank by hand to check for binding. If binding occurs, loosen all caps and tighten individually to determine which bearing is at fault. Check and adjust crankshaft endplay.

Spun Bearing(s)/Bearing Seizure

A lubrication-related failure is caused by insufficient or complete absence of oil in one or all the crank journals. The bearing is subjected to high friction loads and surface welds itself to the affected crank journal. This may result in a spun bearing in which the bearing friction welds itself to the journal and rotates with the journal in the bore or alternatively continues to scuff the journal to destruction. When a crankshaft fractures as a result of bearing seizure, the surface of the journal is destroyed by excessive heat and it fails because it is unable to sustain the torsional loading.

Causes of spun bearings and bearing seizure include:

- Misaligned bearing shell oil hole.
- Improper bearing to journal clearance. Excessive clearance will result in excessive bearing oil throw-off, starving journals furthest from the supply of oil. Insufficient bearing clearance can be caused by over torquing, use of oversize bearing shells where a standard specification is required, and cylinder block line bore irregularities.
- Sludged lubricating oil causing restrictions in oil passages.
- Contaminated engine oil. Fuel or coolant in lube oil will destroy its lubricity.

Etched main bearings are caused by the chemical action of contaminated engine lubricant. Chemical contamination of engine oil by fuel, coolant, or sulfur compounds can result in high acidity levels, which can corrode all metals. However, the condition is usually first noticed in engine main bearings. It may result from extending oil change intervals beyond the recommended specification. Etching appears initially as uneven, erosion pock marks, or channels.

Task C10   Inspect, reinstall, and time the drive gear train (check timing sensors, gear wear and backlash of crankshaft, camshaft, balance shaft, auxiliary drive, and idler gears); service shafts, bushings, and bearings.

The timing gear train includes all gears that drive the camshaft and the assembly must be inspected for damage and wear. The teeth to the gears must be in good shape to maintain the timing of intake and exhaust valves and on some engines, the injectors. Excess tooth wear can affect engine performance and long-term operation and may cause damage to the piston, cylinder, and crank. Carefully inspect for a slight roll or lip on each gear tooth, in addition to looking for chipping, pitting, and burring. If wear is present on any gear, replace it and examine the mating gears for similar damage. Measure backlash using a dial indicator and compare to specifications. Try to determine the cause of damage. If removal is necessary, generally a press or puller will be required. Ensure the gear is properly supported on the center of the hub next to the shaft to prevent cracking or breaking and protect the end of the camshaft. For removal of intermediate and crankshaft gears, refer to the manufacturer's service manuals. Installation of these gears is very critical as it will affect the timing of the engine. The camshaft is generally keyed the same between the shaft and gear, although some manufacturers use offset keys to either retard or advance the timing.

Task C11   Clean, inspect, measure, reinstall or replace pistons, pins, and retainers.

After removing each piston, carefully inspect it and its matching cylinder for possible signs of damage and try to determine probable cause before cleaning. Check for scoring in the skirt of the piston and compare it to the matching side(s) of the cylinder. Scoring could be caused by engine overheating, excessive fuel, improper piston clearance, insufficient lubrication, or a leaking injector or nozzle. A crack in the skirt or lands may be accompanied by bluing or gouges in the skirt and/or cylinder matching with the damage. This can be the result of excessive use of starting fluid or introduction of starter fluid at operating temperatures, excessive piston clearance, or foreign objects in the cylinder like a dropped valve. A nick in the bottom of the skirt and connecting rod, or the cylinder and connecting rod may be caused by improperly installed connecting rods. Uneven wear in the skirt matching a bluing or scuffing on the cylinder could be caused by normal operation at low temperatures, dirty lubricating oil, too little piston clearance, or dirty intake air. The effects of stuck or broken rings may appear on the cylinder as scrapes and gouges, scratches and carbon build-up, or glazing. The leading causes are overheating, insufficient lubrication, and excessive fuel settings. Worn piston pin bores can also be caused by normal wear, insufficient lubrication, or dirty lubricating oil. Under normal conditions, during an overhaul the piston and rings are replaced.

Piston fit problems include:

- Excessive piston to bore clearance results in piston skirt damage and the piston knocking against cylinder wall. The latter is especially noticeable when the engine is cold.

- Too little piston clearance causes piston scoring and scuffing (localized welding). The film of lube oil on the cylinder wall is scraped off.

Piston and Rings Inspection and Failure Analysis

When clamping a piston/connecting rod assembly in a vice, use brass jaws or a generous wrapping of rags around the connecting rod as the slightest nick or abrasion may cause a stress point from which a failure could develop. When attempting to diagnose the cause of a piston failure, use the following a very general guideline:

- Skirt scoring causes overheating, overfueling, improper piston clearance, insufficient lubrication, a leaking injector nozzle.

- A cracked skirt causes excessive use of ether, excessive piston-to-bore clearance, foreign objects in cylinder.

- Uneven skirt wear causes dirt in lube oil, abrasives in air charge, too little piston-to-bore clearance.

- Broken ring lands cause excessive use of ether, excessive piston-to-bore clearance, foreign objects in cylinder.

- Worn piston pin bosses causes old age (normal wear), contaminated lube oil, insufficient lubrication.

- A burned or eroded center crown causes plugged nozzle orifice, injector dribble, cold loading of engine, retarded injection timing, water/coolant leakage into cylinder.

Fuel Injection/Ignition Timing-Related Failures of Pistons

Advanced engine timing causes excessive combustion pressures and temperatures. It may result in torching or blowing out the lower ring lands in severe cases. In less severe cases, erosion and burning of the top ring land or headland area result, as evidenced by pitting.

The latter condition is more common and is often the result of abuse by technicians who believe it will increase engine power. It often does, but at a heavy cost: a significant reduction in engine longevity. Total failure will usually occur in one cylinder with the evidence in the remaining cylinders.

Retarded Engine Timing

Retarded engine timing will cause excessive cylinder temperatures and burning/erosion damage through the central crown area of the piston. An indicator of retarded injection timing is piston crown scorching under the injector nozzle orifice. The condition is seen much less frequently than advanced injection timing because it is seldom intentional and usually the result of technician error or component failure.

Torched Pistons

This term is used to describe a piston or set of pistons that have been overheated to such an extent that meltdown has occurred. Diagnosing a torched piston in isolation from the engine from which it was removed may be a game of guesswork rather than sound failure analysis practice. However, the cause of the condition must be unmistakably diagnosed before the engine is reassembled or a recurrence of the failure will result. If only one of a set of pistons is affected, the diagnosis may be simpler but it is always important to remember that piston torching may be related to lubrication, cooling system, or fuel injection causes.

Piston Pin Retainers

Pistons with full-floating wrist pins have a retainer at each end of the pin. Carefully examine the retaining ring groove area for signs of wear. This area may be eroded due to the excessive temperatures of improperly advanced engine timing. When installing the retaining rings, the open end of the ring should be facing down and *never* at 90° to the connecting rod.

Task C12   Measure piston-to-cylinder wall clearances.

Use a micrometer to measure the piston diameter at right angles to the piston pin bore and 1 in. below the bottom edge of the lowest ring groove. Compare these measurements to the bore diameter of the parent bore. The difference between these two values becomes the piston-to-cylinder wall clearance. The average running clearance on larger engines is 0.006 in. (0.152 mm), but not less than 0.002 in. (0.050 mm). Insufficient clearance will cause premature piston or cylinder/liner failure. Measure cam ground pistons at right angles (90 degrees) to the piston pin. The running clearance of all pistons can be measured using a spring scale with a feeler gauge attached to its end. Insert the 0.006 in. (0.152 mm) feeler gauge in the cylinder (liner). Lubricate the piston with oil and install it (with no rings) bottom up, with the feeler gauge between the cylinder and the piston. Position the piston about 2 in. (50 mm) below the block deck with the piston pin bore in line with the crankshaft. When the spring scale indicates the specified force in pounds as the feeler gauge is being withdrawn, the running clearance is correct.

Task C13   Check ring-to-groove clearance and end gaps; install piston rings.

Ensure the rings have the proper end gap before installing them on the piston. Take this measurement with the rings in the cylinder. Slide each ring into the cylinder, level using a ringless piston, and measure the gap using a feeler gauge. If the end gap clearance is less than specification, the rings may be filed to obtain proper gap. Remove rings from the cylinder and follow manufacturer's instructions for installing these specific rings on the piston. When installing the piston in the cylinder, ensure the ring gaps are spaced according to manufacturer's specifications. Ring side clearance is the installed clearance between the ring and the groove into which it is fitted. The dimension is measured using thickness (feeler) gauges.

Task C14   Identify piston, connecting rod bearing, and main bearing wear patterns that indicate connecting rod and crankshaft alignment or bearing bore problems; check bearing bore and bushing condition; determine needed repairs.

The reasons for piston and engine bearing failures are usually evident in the appearance of the component. When all of the main bearings have been removed from the engine, line up all of the upper bearing shells in order and all of the lower bearing shell in order. The wear patterns should be uniform when compared with each other. Excessive wear in one area is an indication of localized stress of those bearings by something like a misalignment of the bearing bores in the block, an out-of-balance condition, or out-of-round journal. Anything that subjects the engine to uneven loads such as, a faulty harmonic balancer, or an out-of-balance or out-of-alignment flywheel, clutch, or torque converter, will result in uneven wear. The engine is designed to withstand heavy loads, provided they are steady, even, and not excessive. Line up the upper and lower rod bearing shells, and make a similar comparison. When misalignment of the main bearing bores is suspected, begin by checking each bearing bore using an inside micrometer for an out-of-round condition. If any are found to be out of round, the block needs to be line bored. Next, check for alignment of the all of the main bearing bores using a master alignment bar. Misalignment of any one bore will require that the block be line bored. Look for signs that the connecting rods may be bent or twisted. The wear pattern on the piston skirt should be slight, with no evidence of scoring and should be centered on the skirt in a vertical line. Bent or twisted rods will result in a wear pattern on the skirt that is diagonal, not vertical.

All engine manufacturers produce failure analysis guides for technicians. These guides use high definition color photography so that technicians can determine the serviceability of engine internal components.

Because of cylinder combustion forces acting on a piston, the slightest misalignment of the connecting rod will create uneven loading on the wrist pin and compromise the lubricant hydrodynamics at the journal or big end of the rod. When this happens, the big end bearing will show evidence of the misalignment.

Obtain some failure analysis guides from the major OEMs and study the appearance of typical piston and bearing failures.

Task C15   Assemble pistons and connecting rods and install in block; check piston height; replace rod bearings and check clearances; check condition, position, and clearance of piston cooling jets (nozzles).

Installation of semifloating piston pins into an aluminum piston can be made easier by preheating the piston in 200°F (93°C) water. Slide the piston pin through bosses in the piston and connecting rod and install retaining rings. Clamp in a soft jawed vise by the connecting rod to support the assembly while installing rings. Ensure the rod number is correct for the cylinder bore in which it is being placed. Protect the rod bearing and journal by covering rod bolts with plastic caps or a rubber hose, and carefully guide the rod onto the journal until the bearing shell seats. Verify that bearing clearances are within specification using a Plastigage and connecting rod. Check that the rod cap numbers match their cylinder. Sequentially torque the rod cap bolts.

Task C16   Inspect and measure crankshaft vibration damper; determine needed repairs.

A vibration damper reduces the amplitude of vibration and adds to the flywheel's mass in establishing rotary inertia. Its primary function is to reduce crankshaft torsional vibration. The damper consists of a damper drive or housing and inertia ring. The housing is coupled to the crankshaft and using springs, rubber, or viscous medium, drives the inertia ring. The objective is to drive the inertia ring at average crankshaft speed. Viscous-type harmonic balancers have become almost universal in truck and bus diesels; the annular housing is hollow and bolted to the crankshaft. Within the hollow housing, the inertia ring is suspended in and driven by silicone gel. The shearing of the viscous fluid film between the drive ring and the inertia ring affects the damping action. Most OEMs recommended replacement of the harmonic balancer at each major overhaul but this is seldom observed due to the expense and to the fact that these components frequently exceed OEM-projected expectations. The consequences of not replacing the damper when scheduled are economic as they may result in a failed crankshaft. The shearing action of the silicone gel produces friction that is released as heat. This leads to eventual breakdown of the silicone gel, a result of prolonged service life or old age. Drive housing damage, another common reason for viscous damper failure, is caused by careless service facility practice in most cases.

Rubber-type vibration dampers are not often observed on today's heavy truck diesel engines, suggesting that they are probably less effective. The rubber type consists of a drive hub bolted to the crankshaft. A rubber ring is bonded both to the drive hub and the inertia ring. The rubber ring acts both as the drive and the damping medium. The inherent elasticity of rubber enables it to function as a damping medium, but the internal friction generates heat, which eventually hardens the rubber and renders it less effective and vulnerable to shear failures.

To ensure that the balance of the crankshaft is maintained during assembly of the engine, mark alignment of all drivetrain components before removal. Inspect the vibration damper, flywheel/ flexplate, and clutch/torque converter for damage that may cause an unbalanced condition. Damage can also be seen in uneven wear on the main bearings and/or stress on the crankshaft. Inspect the flywheel for cracks, missing or damaged ring gear teeth, distortion, oblong mounting holes, or a pilot bearing with a loose or improper fit.

Visually inspect the damper housing, noting any dents or signs of warpage. Evidence of either is reason to reject the component. Use the following process:

- Using a dial indicator, rotate the engine manually and check for damper housing radial and axial runout against the OEM specification. This is a low tolerance specification usually 0.005 in. (0.127 mm) or less.

- Check for indications of fluid leakage, initially with the damper in place. Trace evidence of leakage justifies replacement of the damper.

- Should the damper not be condemned using the above tests, remove it from the engine. By hand, shake the damper. Any clunking or rattle is reason to replace it.

- Next, using a gear hotplate or component heating oven, heat the damper to its operating temperature, usually close to the engine operating temperature—around 90°C (180°F). This may produce evidence of a leak.

- A final strategy is to mount the damper in a lathe using a suitable mandrel. Run it up through the engine operating range and monitor it using a balance sensor and strobe light.

- While on the crankshaft check for wobble using a dial indicator; replace it if it is not within the manufacturer's specifications. Viscous dampers should be checked for nicks, cracks, or bulges. Cracks or bulges may indicate that internal fluid has ignited and caused the case to expand.

Task C17   Inspect, install, and align flywheel housing.

Whenever a flywheel housing is removed from the cylinder block, the technician must check the housing innerflange face concentricity with the crankshaft using a dial indicator. The maximum tolerance for the crankshaft to flywheel housing eccentricity is low, typically around 0.012 in. (0.3 mm) total indicated reading (TIR) and the consequences of installing a flywheel housing that exceeds the allowable specification are severe. When flywheel eccentricity is present, the drive axis is broken. This condition can result in clutch, engine mount, transmission, and engine failures. The flywheel housing-to-crankshaft concentricity is preserved by interference fit cylindrical or diamonds dowels. These dowels may be relied upon to properly realign the flywheel housing on the cylinder block at each reinstallation, but the specification is so critical it should be checked. The following procedure outlines a method of checking flywheel housing concentricity and typical strategies for correcting an out-of-specification condition.

1. Ensure that the engine is properly supported. Indicating flywheel housing may be performed with the engine in or out of the chassis. Locate the flywheel innerflange face to crankshaft concentricity TIR tolerance in the OEM service manual. Either mechanically or electrically ensure the engine fuel system will not function so the engine will not start.

2. Mount the flywheel housing to the engine cylinder block using the dowels to align the assembly and snug the flywheel fasteners at about half of the OEM-specified torque value.

3. Use fabricator's chalk to mark the flywheel flange face in four evenly spaced locations, such as at N, E, S, and W or NW, NE, SE, and SW.

4. Next, thoroughly clean the inside face of the flywheel housing with emery cloth and fix a magnetic base dial indicator to any position on the crankshaft, setting the plunger to contact the flywheel housing inside face. Using an engine barring tool, rotate the engine in its normal direction of rotation until the indicator probe is positioned at any one of the chalk indexes. Now set the indicator to 0.

5. Bar the engine through a full rotation, recording the indicator reading at each of the indexes. The indicator should once again read 0 when the revolution is complete. If the readings are A = 0, B = –0.003 in., C = –0.005 in., and D = +0.004 in., the TIR would be the highest negative value (–0.005 in.) added to the highest positive value (0.004 in.), giving a reading of 0.009 in. If the OEM TIR maximum specification was 0.012 in., this reading would be within it. If the flywheel housing-to-crankshaft concentricity is outside of the specification, it can be reset.

Task C18   Inspect flywheel or flexplate (including ring gear) and mounting surfaces for cracks, wear, and runout; determine needed repairs.

Checking Flywheel Runout

Mount a dial indicator with a magnetic base on the flywheel housing. Check the flywheel runout at the clutch contact face. The typical allowable runout is 0.001 in. per inch of clutch radius (0.001 mm per millimeter of clutch radius). Measure the radius from the center of the flywheel to the outer edge of the clutch contact face. For example, a 14-in. (355 mm) clutch would have a 7-in. (177 mm) radius and allow a 0.007 in. (0.177 mm) total indicator reading of the dial indicator. If the measured values fall out of these specifications, repeat the assembly process and remeasure until a cause can be determined.

Flywheels are commonly removed from engines for reasons such as clutch damage, leaking rear main seals, leaking cam plugs etc., and care should be taken both when inspecting and reinstalling the flywheel and the flywheel housing. Flywheels should be inspected for face warpage, heat checks, scoring, intermediate drive lug alignment and integrity, axial and radial runout using dial indicators, straightedges, and thickness gauges. Damaged flywheel faces may be machined using a flywheel resurfacing lathe to OEM tolerances: typical maximum machining tolerances range from 0.060 to 0.090 in. (1.50 mm to 2.30 mm). *Note:* When resurfacing pot-type flywheel faces, the pot face must have the same amount of material ground away as the flywheel face. The consequence of machining the clutch face only will result in an inoperable clutch.

Inspect the flywheel for cracks, missing or damaged ring gear teeth, distortion, oblong mounting holes, or a pilot bearing with a loose or improper fit. Inspect the vibration damper, flywheel/flexplate, and clutch/torque converter for damage that may cause an unbalanced condition. If damage is detected, check for uneven wear on the main bearings and/or stress on the crankshaft as well.

Shrunkfit to the outer periphery of the flywheel is the ring gear, which is the means of transmitting cranking torque to the engine by the starter motor during startup.

Vehicles equipped with automatic transmissions use flexplates in place of the flywheel assembly. Flexplates provide the connection between the engine crankshaft and the transmission torque converter. Generally, the flexplate assembly consists of 4 or 5 thin steel plates held to the crankshaft with a retainer plate and bolts. The plates attach to the outer periphery of the torque converter by bolts accessed through a hole in the flywheel housing. Starter ring gears are usually mounted to the torque converter directly. Flexplate assemblies must be inspected for cracks, particularly around the mounting bolt holes. Flexplate failures are caused by loose transmission to engine mounting bolts, worn front support bearing in the transmission, and improper engine to transmission alignment.

D. Lubrication and Cooling Systems Diagnosis and Repair (8 Questions)

Task D1 **Verify engine oil pressure and check operation of pressure sensor, pressure gauge, and sending unit.**

Oil pressure gauges are either electrically or mechanically operated and low oil pressure lights are electrically operated indicators. These devices must be considered any time incorrect or faulty oil pressure readings are obtained. Usually mechanical gauges are reliable and seldom fail.

Mechanical gauges use the engine oil pressure routed through a tube to operate the meter movements. If the tube becomes punctured, chinked, or clogged, the fluid will not be able to operate the meter movements. In this case, the tube must be replaced if it is punctured or pinched, or blown clean if clogged.

Common problems associated with electrical gauges are open or shorted wiring or circuitry, clogged passages to the sensor, or a defective sensor. The gauge can become damaged if too much voltage is applied, but this usually is a symptom of an electrical wiring problem. Use a digital multimeter (DMM) to check the resistance of the sensor and continuity of the wiring. If the sensor is at fault, replace it and check the oil passage for obstructions. If the wiring is at fault, repair or replace it and determine the cause.

Indicator lights use a method similar to the electrical gauge except they incorporate a pressure switch and a light bulb.

The newest dash assemblies are provided information on the data bus. The engine oil pressure information is processed by the engine ECM and sent to the instrument cluster on the data bus.

Task D2 **Inspect, measure, repair or replace oil pump, drives, pipes, and screens.**

When removing and inspecting engine oil pumps, refer to the manufacturer's manual for specific instructions and required tools. Some manufacturers use external oil pumps that have oil lines routed out of the block and return near the by-pass valve. Most use internal oil pumps that can be accessed from the crankcase by dropping the oil pan with conventional tools. Remove the bolts that mount the oil pump to the engine and then carefully remove the oil pump. With external oil pumps, remove the cover from the pump body and inspect it for wear. Remove and inspect the gears for pitting and wear. Check the diameter of the gears and driveshaft using a micrometer and compare with manufacturer's specifications. Check all mating surfaces and the pump body for damage and wear. After reinstalling the gears and driveshaft in the body, check the gear-to-cover clearance with Plastigage. If it fails to meet manufacturer's specifications, replace the worn parts. If the pump has an integral bypass valve, it must also be checked for wear and cleaned. Because oil pumps are well lubricated by the fluid they pump, there will be little wear unless they are damaged by foreign particles passing through. An engine that has been well maintained should have an oil pump showing very little wear.

Task D3　　**Inspect, repair or replace oil pressure regulator valve(s), by-pass valve(s), and filters.**

Oil filters trap particles that would cause damage to critical components, such as engine bearings. Regular replacement of filter elements and inspection of the housing is necessary. Engines use a variety of materials from cast iron to die cast aluminum for the housing and steel or plastic for machined parts. For this reason, the technician is often limited to visual inspection of filter assembly parts. Cracks anywhere in the housing and nicks on machined surfaces can cause engine failures. Some filter assemblies incorporate a by-pass valve in their housing. These valve assemblies must be removed and inspected for corrosion, wear, and other signs of damage during overhaul or any time the filter assembly contains metal particles.

Canister filters can be cut open and the element material examined for content. Metal particles indicate a potential bearing failure. A milky gray sludge indicates water in the oil caused by a possible head or cylinder block leak, which may already have caused bearing damage. Ensure all gasket surfaces are straight and free of nicks, which could cause an improper seal of the assembly. When inspecting element-type filter assemblies, ensure that housing grooves, center bolt threads, springs, metal gaskets, and spacers are in good condition. Inspect for distortions in the canister, cracks around canister bolt hole and clogged passageways in the bolt. Reinstall the by-pass valve according to manufacturer's installation procedures and set for proper trip pressure if adjustable. Replace the element and all gaskets and seals. When reassembling filter assemblies, apply a small amount of oil to seals to lubricate them. Priming of filters is recommended to ensure adequate oil is present during startup to lubricate the bearings.

Task D4　　**Inspect, clean, test, reinstall or replace oil cooler, by-pass valve, oil thermostat, lines and hoses.**

Complete servicing of the oil cooler should be done as part of a major engine overhaul, or whenever there is oil in the water, water in the oil pan, or a milky gray sludge in the filter. In any of these cases, specific manufacturer's instructions will apply. Look for places where the oil cooler baffles may have vibrated against the tubes. This common failure results in leaks. If you find signs of this, replace the cooler. Look for gasket particles and debris that are obstructing fluid flow and clean thoroughly. Pressure test the oil cooler core for leaks before reuse. The cooler by-pass valve should have a slightly worn appearance and have no signs of scoring. Check the lines and hoses for signs of leaking. With a small light, inspect the inside of hoses, since they deteriorate from the inside first. Replace any hoses with internal cracks.

Task D5　　**Inspect turbocharger lubrication system; repair or replace as needed.**

To check the turbocharger lubrication, begin by looking for signs of leaks. Inspect hoses for fraying or cutting and tubes for kinks or nicks. Remove hoses and tubes and check for obstructions. Install the lines after replacing any damaged ones and install a tee in one of the connections. Attach an oil pressure gauge to the tee and tighten all connections. Start the engine and monitor the oil pressure gauge. Operate the engine over a range of operating speeds while monitoring the oil pressure. Stop the engine and ensure the pressure drops. All gauge readings must mimic the instrument panel oil pressure gauge. Ensure that the oil return line is not obstructed. Restriction to flow on the return will cause the oil to remain in the bearing housing too long, resulting in high oil temperature. This condition can lead to premature bearing failure and coking of the oil in the bearing housing.

Task D6　　**Change oil and filter, verify oil level and condition.**

Oil change procedures, lube oil and filter requirements, and oil level checking procedures may vary from one engine manufacturer to another. Some manufacturers require prefilling the filter at installation, while others require the filter to be installed empty to avoid unfiltered oil from entering the lube system. Viscosity requirements, API ratings and additive packages must be taken into consideration when selecting oil. If the wrong oil is used, serious engine damage can result. Newer engines equipped with EGR controls require low ash oils to comply with emissions ratings. Supplemental or by-pass oil filtration systems are also commonly used in transit applications, such as "Spinner II" oil cleaning centrifuges. These systems commonly use centrifugal force to remove contaminates from the oil. These filters will remove much smaller particles (as small as 10 microns) from the oil since they remove contaminates by weight and not by size. Centrifugal filters trap impurities in a bowl that needs to be serviced along with the full-flow filters at the normal oil change intervals.

Task D7   Inspect, reinstall or replace drive belts, pulleys and tensioners; adjust drive belts and check alignment.

To install drive belts:

- When two or more identical belts are used on the same pulley, all of the belts must be replaced at the same time.
- Make sure the distance between the pulley centers is as short as possible when you install the belts. Do not roll the belts over the pulley. Do not use a tool to pry the belts onto the pulley.
- The pulleys must not be out of alignment more than 1/16 in. (1.59 mm) for each 12 in. (30.5 mm) of distance between the pulley center. Pulley alignment may be checked with a straightedge.
- The belts must not touch the bottom of the pulley grooves, and they must not protrude more than 3/32 in. (2.38 mm) above the outside diameter of the pulley.
- When identical belts are installed on a pulley, the protrusion of the belts must not vary more than 1/16 in. (1.59 mm).
- Make sure that the belts do not touch or hit any part of the engine.

To adjust drive belts:

- Use the belt tension gauge to check the tension of the belts.
- Tighten the reading of 110 to 160 lbs. as indicated on the gauge.
- After the engine has been running for at least 1 hour, stop the engine and check the belt tension. If the tension is less than the value given in the previous step, adjust the belt to the correct value.

Many late-model buses use a serpentine belt drive with V-ribbed, or multi-V style belts. They are called serpentine because the belt follows a snake-like path around the pulleys. The belt is thinner and more flexible so that it can bend around smaller pulleys and can also be bent backwards, allowing the use of both sides of the belt to transfer power. One belt usually drives all of the belt-driven accessories, saving space, and transferring power more efficiently. They also last longer than conventional V belts, and are easier to replace. A spring-loaded tensioner and idler pulleys are usually used to apply constant belt tension. Other methods of tensioning the belt include adjustable jackscrews and a tensioner with an off-center bolt. Examine the underside of the belt. If 10 or more cracks per inch are present, replace the belt. The idler and tensioner pulleys should be checked for noisy, dry, or worn bearings, and for wear to the pulley surface. If V-ribbed belts are noisy or walk off the pulleys, look for misalignment of the pulleys.

Task D8   Verify coolant temperature, and check operation of temperature and level sensors, temperature gauge, and sending unit.

Coolant temperature is used by electronically controlled diesels to monitor and control the engine operation as well as to keep the driver informed. Standard temperature gauges and indicator lights fall into one of two types: mechanical or electrical. Mechanical gauges use a liquid-filled sensor that functions much like a thermometer. This gauge is self-contained and must be replaced as an assembly. Common reasons for failure of a mechanical gauge are corrosion on sensor, leaking sensor, pinched tube, and binding of meter movement of needle. Electrical gauges use a sensor known as a thermister that changes its resistance as temperature changes. This resistance is used to vary a return signal voltage value. Failures of an electrical gauge can be caused by defective sensors, an open or short in wiring, and open or bad grounds.

Indicator lights typically use a sensor that acts like a switch to turn on a light circuit. Probable causes for failure are the same as an electrical gauge except the lightbulb itself can fail. With the introduction of electronics into the management of diesel engines, sensors are no longer directly connected to display gauges. As such, a problem may be with a faulty module or circuit card. Other problems may be introduced by software programs or scrambled input circuit signals.

Task D9 **Inspect and replace cooling system thermostat(s), by-passes, housing(s), and seals.**

To function effectively a thermostat must:

- start to open at a specified temperature,
- be fully open at a set number of degrees above the start to open temperature,
- define a flow area through the thermostat in the fully open position, and
- permit zero coolant flow or a defined small quantity of flow when in the fully closed position.

The cooling system thermostat is normally located either in the coolant manifold or in a housing attached to the coolant manifold. Its primary function is to permit a rapid warmup of the engine. When the engine has attained its normal operating temperature, the thermostat will open and permit coolant circulation. As the thermostat defines the flow area for circulating the coolant, there may be more than one. A heat-sensing element actuates a piston that is attached to the seal cylinder. When the engine is cold, coolant is routed to the coolant pump to be recirculated through the engine. When the engine heats to operating temperature, the seal cylinder gates off the passage to the coolant pump and routes the coolant to the radiator. The heat-sensing element consists of a hydrocarbon or wax pellet into which the actuating shaft of the thermostat is immersed. As the hydrocarbon or wax medium expands, the actuating shaft is forced outward in the pellet, opening the thermostat. Thermostats can be full blocking or partial blocking.

Top By-pass Thermostat

The top by-pass thermostat simultaneously controls the flow of coolant to the radiator and the by-pass circuit. During engine warmup, all of the engine coolant is directed to flow through the by-pass circuit. As the temperature rises to operating temperature, the thermostat begins to open and coolant flow is routed to the radiator, increasing incrementally with temperature rise.

Poppet or Choke Thermostats

Poppet-type thermostats control the flow of coolant to the radiator only and the by-pass circuit is open continuously. Flow to the radiator is discharged through the top of the thermostat valve.

Side By-pass or Partial Blocking Thermostat

The side by-pass thermostat functions similarly to the poppet type. It has a circular sleeve below the valve that moves with the valve as it opens. This serves to partially block the by-pass circuit and direct most of the flow to the radiator.

Vented and Unvented Thermostats

Vented thermostats have a small orifice in the valve itself or a notch in the seat. Usually this must be positioned in a upright position on installation. The vent orifice helps to deaerate the coolant by routing air bubbles out of the by-pass circuit. Positive deaeration-type systems usually require nonvented thermostats.

By-pass Circuit

The term by-pass circuit describes the routing of the coolant before the thermostat opens; that is, through the engine cylinder block and head. The flow of by-pass coolant permits rapid engine warmup to the required operating temperature.

Running without a thermostat is not recommended. It also contravenes EPA requirements regarding tampering with emission control components. Removing the thermostat invariably results in the engine running too cool, causing vaporized water in the crankcase to condense and producing corrosive acids (sulfuric acid) and sludge in the crankcase. Additionally, low engine running temperatures will increase the emission of HC. Conversely, engines that should use top by-pass or partial by-pass-type thermostats may overheat when the thermostat is removed as most of the coolant will be routed through the by-pass circuit with little being routed in the radiator. Remember, there is a difference between start to open and fully open temperature values.

Task D10   Flush and refill cooling system; bleed air from system; recover coolant.

Engine coolant is a mixture of water, antifreeze, and supplemental cooling additives (SCA). To cool the engine, coolant must have an unobstructed flow throughout the cooling system. Often when refilling the cooling system, air becomes trapped inside the block. If this condition persists and the engine temperature exceeds 212°F (100°C), water will turn to steam. Steam pressure forces more water out of the block, causing a gas lock in the cooling system. To prevent this from happening, manufacturers usually install bleeder valves or vent lines and provide instructions on their use. Generally, bleeder valves are located near the top of the cooling system and block. After a cooling system has been in operation for several years, its efficiency may be reduced due to formation of scale and sludge. If such formation occurs, the radiator may need to be removed and flushed. While this can be performed in the vehicle, it generally is not very effective. Drain the coolant system into a container by opening the lower drain valves and upper bleed valves. Coolant should be tested before reusing. When coolant must be replaced, it is a good practice to premix the coolant in a container before filling the system. Once the radiator or some other component is replaced, refill the system with the bleeder valves open. When all the air is out of the system, close the bleeder valves, start the engine, and check the coolant level. Test the antifreeze protection level at operating temperature. Antifreeze will prevent corrosion and raise the temperature at which the coolant will boil under pressure. This procedure lengthens the life of the engine and cooling system components.

Task D11   Inspect, repair or replace coolant conditioner/filter, check valves, lines, shutoff valves, and fittings.

When freeze protection is required, an antifreeze meeting engine manufacturer's specification must be used. An inhibitor system is included in most types of antifreeze and no additional inhibitors are required on initial fill if a minimum antifreeze concentration of 30 percent by volume is used. Solutions of less than 30 percent concentration do not provide sufficient corrosion protection. Concentrations over 67 percent adversely affect freeze protection and heat transfer rates.

Supplemental Coolant Additives (SCA)

Supplemental coolant additives are critical to the coolant mixture. The actual SCA package recommended by an engine manufacturer will depend on whether wet or dry cylinder liners are used, the materials used in the cooling system components, and the fluid dynamics (high flow/low flow) within the cooling system. The operator may want to adjust the SCA package to suit a specific operating environment or set of conditions. For instance, abnormally hard water will require a greater degree of antiscale protection. Depending on the manufacturer, SCA may be added to a cooling system in a number of ways. When there are excessive additives, the system must be drained and refilled with the proper mixture. If there is only slight overconditioning, some manufacturers allow continued operation until the next PMI. Most OEMs suggest testing the SCA levels in the coolant followed by adding SCA to adjust to the required values. Never dump unmeasured quantities of SCA into the cooling system at each PM.

Testing SCA Levels

Generally OEMs recommend that the coolant SCA level be tested at each oil change interval. Additionally, whenever a substantial loss of coolant has occurred and the system has to be replenished, the SCA level should be tested. Each test provided by an OEM is designed to monitor the SCA package required for its product and cannot generally be used for other OEM products. Also, test kits usually consist of test strips, which must be stored in airtight containers and have expiration dates that should be observed. Coolant test kits permit the technician to test for the appropriate SCA concentration, the pH level, and the total dissolved solids (TDS). The pH level determines the relative acidity or alkalinity of the coolant. Acids may form in engine coolant exposed to combustion gases or in some cases when cooling system metals (ferrous and copper base) degrade. The pH test is a litmus test in which a test strip is first inserted into a sample of the coolant, then removed and the color of the test strip indexed to a color chart provided with the kit. The optimum pH window is defined by each OEM, but normally falls between 7.5 and 11.0 on the pH scale. Higher acidity readings (below 7.5 on the pH scale) in tested coolant are indications of corrosion of ferrous and copper metals, coolant exposure to combustion gases, and, in some cases, coolant degradation. Higher than normal alkalinity readings indicate aluminum corrosion and possibly that a low silicate antifreeze is being used where a high silicate antifreeze is required.

Task D12 Inspect, repair, or replace water pump, housing, hoses, idler pulley and drives.

Most water pumps are belt driven, using an idler pulley to maintain belt tension. As belts age they stretch and the idler pulley (on most engines) automatically adjusts tension. If the belt is loose, check for binding in the idler pulley. Water pumps usually allow a small amount of coolant to weep around the bearings to cool and lubricate them. Expect a few drops to leak from a factory hole in the bottom of the pump casting. Most water pumps are mounted to standoffs on the engine and require inlet and outlet hoses. More manufacturers are mounting them into the front of the block and some require no hose connections. With this approach a gasket is used between the block and water pump. While any damage to a belt is reason to change it, most manufacturers say it must be changed if two or more cracks appear in any 2-in. length.

Task D13 Inspect radiator, pressure cap, and tank(s); determine needed service.

The radiator has a pressure-control cap with a normally closed valve. A cap, with a 7 stamped on its top, is designed to permit a pressure of approximately 7 psi (48 kPa) in the system before the valve opens while the cap with a 9 stamped on its top needs 9 psi (62 kPa) in the system before the valve opens. Typical maximum system pressures seldom exceed 15 psi. This maximum pressure raises the boiling point of the cooling liquid and permits somewhat higher engine operating temperatures without loss of any coolant from boiling. To prevent the collapse of hoses, a second valve in the cap opens under vacuum when the system cools, or when there is about 0.25 psi pressure difference between coolant and atmospheric pressure. Use extreme care when removing the coolant pressure-control cap. Remove the cap slowly after the engine has cooled. The sudden release of pressure from a heated cooling system can result in loss of coolant and possible personal injury (scalding) from the hot liquid. To ensure against possible damage to the cooling system from either excessive pressure or vacuum, check both valves periodically for proper opening and closing pressures. If the pressure valve or the vacuum valve does not open at the specified pressures, replace the pressure control cap.

Task D14 Inspect, repair, or replace fan hub, fan, and fan clutch; inspect mechanical, hydraulic, and electronic fan controls, fan thermostat, and fan shroud.

Fan assemblies must be inspected. Clean the fan and related parts with soap and water and dry them with compressed air. Shielded bearings must not be washed as dirt may be washed in. Examine the bearings for corrosion or pitting. Hold the inner race or cone so it does not turn and revolve the outer race or cup slowly by hand. If rough spots are found, replace the bearings. Check the fan blades for cracks. Replace the fan if the blades are bent, since straightening may weaken the blades, particularly in the hub area. Remove any rust or rough spots in the grooves of the fan pulley and crankshaft pulley. If the grooves are damaged or severely worn, replace the pulley. Many transit buses utilize hydraulically driven fans, of which most are multispeed. Depending on operating conditions, the fan may be not rotating, or rotating at low, medium, or high speeds. Inspection includes, but is not limited to, checking the mounting of the fan motor, leaks at the fan motor, lines and controls, hydraulic fluid level and condition, and proper operation of fan speed controls. Refer to OEM service information for troubleshooting procedures of fan speed controls. Modern buses have computer-controlled fans that may be turned on by the engine ECM, the air conditioning, or the engine brake. Examination of fan controls must include verifying the operation of computer input sensors, such as the engine coolant temperature (ECT) sensor, inlet air temperature (IAT) sensor, and the engine oil temperature (EOT) sensor. Check the engine cooling fan solenoid, which is normally closed, and, when energized by the ECM, supplies air pressure to disengage the engine cooling fan clutch. The engine cooling fan manual switch is a normally open switch that, when closed by the operator, requests that the ECM deenergize the engine cooling fan solenoid. This causes the engine cooling fan to operate continuously. Use the OEM recommended electronic service tools (EST) to verify operation of ECM-controlled components. Other systems use fan thermostats (fanstats) with a temperature-sensing bulb located in the water jacket, which either opens or closes depending on temperature and system design. Some fan clutches are spring-loaded to the engaged position and are released by air pressure; other designs are opposite of this. Be sure to check the service manual for the correct procedure.

Task D15   Pressure test cooling system and radiator cap; determine needed repairs.

Cooling system leaks are a common problem, so the operator should check the cooling system daily. Whenever it is necessary to replenish coolant frequently, a pressure test should be performed to locate the source of the loss. Pressure testing will easily locate most external leaks. Leaks are usually located by a good visual inspection, since antifreeze leaves a film wherever there is a leak. Cold leaks occur due to contraction of sealing components, especially hose clamps. They often cease to leak at operating temperatures. Silicone hoses are frequently used due to their longer service life. Always use the special clamps they require, and be careful to torque them properly since they are sensitive to overtightening. A cooling system pressure testing kit includes a hand-actuated pump and gauge, and the necessary adapters to fit different radiators and pressure caps. Attach the pump using the necessary adapters to the radiator and pump up pressure to the pressure range of the cooling system. Observe the system's ability to hold pressure while looking for visible signs of leakage. The radiator cap may also be tested using the pump and gauge assembly. Radiator caps are rated by the pressure to overcome the cap's spring pressure and unseat the seal. When this occurs coolant is routed to the surge tank. When the cooling system cools after the engine is shut down, pressure drops. When it falls to a point where atmospheric pressure is about 0.25 psi higher than the coolant pressure, a vacuum valve in the cap unseats and coolant is forced by atmospheric pressure from the surge tank back to the radiator. When this valve fails to open, radiator hoses may be "sucked" flat. Internal leaks are harder to find and are usually located because coolant is found in places where it does not belong, such as in the oil, fuel, or exhaust.

E. Air Induction and Exhaust Systems Diagnosis and Repair (8 Questions)

Task E1   Inspect, service or replace air induction piping, air cleaner, and element; check for air restriction or contamination.

Most air filters are designed to route air to circulate around the filter element. This cyclonic action causes airborne particles (like sand and dust) to be slung against the sides of the canister where they fall and collect in the bottom chamber, reducing the clogging of the filter element and extending its life. The filter element is impregnated with a resin to trap most of the remaining particles. This type of air filter is the most common and has about a 99.6–99.9 percent efficiency rating.

Excessive restriction of the air inlet will affect the flow of air to the cylinders and result in poor combustion and lack of power. A normal maximum spec is 25 inches of water at full load. Consequently, inlet restriction must be kept within specifications. An obstruction in the air inlet system or dirty air cleaners will result in a high inlet restriction. Check inlet restriction with a water manometer connected to a removable fitting in the air inlet ducting (nonturbocharged engine) or the compressor inlet (turbocharged engine). Check the normal air inlet vacuum (restriction) and compare the results with the manufacturer's recommended specifications. Most service manuals require the engine to be operated at full load while measuring the restriction. This method allows the turbo to be at full boost, therefore the maximum amount of air is flowing through the air filter. Others require the restriction to be measured at high idle (no load).

Flow pulsation can occur in the air intake when the engine is equipped with an engine brake. This is caused by cylinder pressures and the intake valves opening during the braking period. This flow reversal and pulsation can damage the element and cause dirt to move through the element and enter the engine. An air intake suppressor can help to prevent this problem. The air intake suppressor must be installed into the air intake system between the engine and the air cleaner. Install the suppressor as close to the engine as possible.

Task E2   Inspect, test, and replace turbocharger, wastegate, and wastegate controls.

Turbochargers are used on almost all bus diesel engines to create manifold boost; that is, they pressurize the air delivered to the cylinders. Exhaust gas-driven turbine blades on the exhaust side of the turbocharger drive compressor blades on the intake side of the compressor. Because the rotor and compressor are connected by a common shaft and built to close tolerances, the assembly can achieve peak speeds between 60,000 to 150,000 rpm. Variable geometry turbochargers develop boost at lower rpms, reducing turbo-lag time. The resultant increase in air density increases engine output. Compressing air in a turbocharger heats it up. To reduce the effects of superheating, a heat exchanger, known as an aftercooler, is located between the turbocharger and the cylinders. While the commonly used aftercooler is an air-to-air heat exchanger, some engine manufacturers prefer to use an air-to-water heat exchanger, called an intercooler. Regardless of the method used, the result is increased efficiency because cool air is denser.

The rotor on the exhaust side of the turbo is called the turbine, because it is driven by exhaust gases. The impeller of the intake side is called the compressor, because it compresses the air charge to the engine.

As exhaust gases drive the turbine, the compressor is rotated. Rotation of the compressor causes air to be drawn from the outside, through the filter and into the inlet side of the compressor. Pressurization causes the air to heat and additional heat is transferred from the turbine through the shaft to the compressor and to the compressed air. This hot pressurized air is routed to the aftercooler where it is cooled and directed to the intake manifold. After combustion, hot expanding gases are routed through the exhaust manifold to the turbine inlet. The hot gases turn the turbine and exit through the outlet to the exhaust and muffler. Some turbochargers use an electrically or mechanically controlled valve to route a portion of the exhaust gases around the turbine. This valve, referred to as a wastegate or overboost protection valve, prevents the turbocharger from overboosting the cylinders at high engine rpm. Modern electronic systems incorporate a boost pressure sensor to monitor the turbocharger and an electronic module to control the wastegate or overboost valve.

When inspecting a turbocharger, begin with a visual inspection. Check the turbocharger outlet for the presence of oil. Oil in the outlet indicates that the turbocharger seals might be leaking. The seals may be worn, or their leakage may be caused by a restricted air cleaner, a turbocharger oil drain that is restricted or improperly routed, or high crankcase pressure. Visually inspect the turbine and compressor wheels for damaged blades. Check the turbocharger shaft for axial and radial endplay and compare to specifications. If damaged blades or endplay that is out-of-spec is found, check for damage to the housings. If the housings are also damaged, the turbocharger should be replaced as an assembly. If the housings are still serviceable, the center housing rotating assembly (core) may be replaced. It is a prebalanced assembly, recoring is merely a matter of transferring the housings and being careful to observe the correct alignment. When recoring or replacing a turbocharger, pour clean oil into the oil supply line fitting to prelube the turbocharger before starting the engine. Some causes of turbocharger failures are: hot shutdowns, which produce warped shafts and bearing failures; turbocharger overspeed from high altitude operation or overfueling; lubrication-related failures; and ingestion of dirt or foreign objects.

Task E3   **Inspect and replace intake manifold and gaskets; test temperature and pressure sensors; check connections.**

Intake piping, tubing, and hoses are durable yet flexible to allow for movement of the engine. To reduce weight, solid piping is usually made of aluminum or ridged plastic. Connections between the solid section of piping and major components are made using rubber sections clamped at both ends. When these connections become loose, unfiltered air is allowed to enter the intake system. Airborne abrasive or corrosive particulates can then cause damage to critical components. On the other hand, as the rubber or plastic components age or become exposed to extreme heat, they will collapse and cause restrictions in the airflow. To provide a visual warning to the operator of problems in the air intake system, most manufacturers include an air intake inlet restriction indicator. This could be as simple as a vacuum tube attached to the intake piping that runs to an instrument panel indicator, or an electronic sensor that is monitored by an electronic module. If the inlet restriction increases, check for a restriction such as a collapsed pipe or clogged filter. A drop in vacuum indicates a possible leak, clogged vacuum indicator tube, or defective sensor.

Two sensors are located on or near the intake manifold. The solid state boost pressure (BP) sensor monitors positive intake manifold pressure, and is used by the ECM to control fuel metering and injection timing and to limit smoke during acceleration. The intake manifold temperature (IMT) sensor monitors intake manifold temperature. The ECM uses this signal to control injection timing, fuel metering, and engine protection.

Task E4   **Inspect, test, clean, repair or replace aftercooler or charge-air cooler and piping system.**

The aftercooler resembles a radiator except the fin tubes appear to be larger and the inlet and outlet connections are larger. This allows a large volume of air to pass through. As the hot pressurized air passes through the small fin tubes, heat is transferred from the pressurized air to the tubes, then to the fins, and finally to the ambient air passing between the fins.

Aftercoolers usually can be inspected visually while still in the vehicle. Wash fins with low-pressure water and straighten as necessary. Inspect for missing or loose fins, dented or cinched tubes, corrosion, or holes in tubes. Leaks can be detected by running the engine, spraying the aftercooler tubes with soap and water, and watching for bubbles.

An intercooler cools the pressurized air using liquid coolant in the opposite way the radiator works. Heated air is passed through small tubes inside a chamber, usually aluminum, through which coolant is circulated. Heat from the air is transferred to the tubes and onto the coolant, which is cooled by the radiator.

Intercoolers are more difficult to inspect, but typically, the indicator is engine performance. If the engine has been shut down for a while, traces of water may appear in the oil or the engine may produce a white smoke when running. The turbocharger may have signs of liquid leakage from its compressor outlet or corrosion on the fins. This would also result in unexplained loss of coolant in the radiator, while the engine is still cold. Often internal leaking of the intercooler can be misdiagnosed as a cracked head or blown head gasket. A common problem with intercoolers is airlocks. When cooling system maintenance is performed on the engine resulting in refilling, the coolant air must be bled from the intercooler coolant chamber. Failure to do so could reduce the effectiveness of the cooler and cause overheating due to steam pressure build-up.

Task E5   **Inspect, repair or replace exhaust manifold, gaskets, piping, mufflers, insulation/heat shield and mounting hardware; inspect, replace, or repair exhaust after treatment devices.**

Most exhaust manifolds are manufactured from cast iron sections flanged to the exhaust tracts of the cylinder heads. Many are tuned; that is, the exhaust manifold is designed to deliver exhaust gas to the turbocharger so that as a pulse of end gas from one cylinder is unloaded into the manifold, the vortex of the pulse before it helps pull it toward the turbine housing of the turbocharger. A tuned or pulse turbo is also designed to optimize the delivery of hot end gas onto the turbine vanes to produce better turbocharger efficiencies.

Most diesel exhaust piping components are manufactured from galvanized steels. Stainless steels and chrome surfacing are also used, especially where appearance is a factor. When leaks occur, they are usually at the clamp joints and are usually detectable by soot tattletales at the point of the leak. When stainless steel band clamps are used, it is good practice to replace them each time the exhaust system is disassembled.

Mufflers in diesel applications use both resonation and sound absorption principles to change the frequency of exhaust noise. The muffler is an emission control device (sonic emissions), so when one is replaced, it must meet the specifications of the original muffler.

Catalytic converters are now more common on diesel engines. Currently, most catalytic converters are single-stage oxidizing converters. Some newer bus mufflers may also contain diesel particulate filters or DPFs, which are designed to allow exhaust gasses to pass through them, but will trap and contain any solid material such as soot and ash. When the soot and ash collect, the exhaust system back pressure will be affected. Back pressure levels must be monitored. When back pressure reaches a predetermined limit, the engine ECM will illuminate the Check Engine Light (CEL) and log a code. This will inform the technician that the DPF needs to be serviced. Most DPF muffler designs allow removal of the DPF separately from the rest of the muffler/catalyst assembly. When catalytic converters are used, it is essential to adhere to the original piping geometry. Any changes to the gas dynamics through the exhaust piping upstream from a catalytic converter can cause the device to overheat or not function at all. Because it is an emission control device, a catalytic converter cannot be removed. If it has to be replaced, the replacement unit must meet the original specifications.

Task E6   Inspect, repair or replace preheater/inlet air heater, starting aids, and controls.

Glow plug circuits on modern diesel engines are controlled by an ECM. A glow plug relay control is used to energize the glow plugs for assisting cold engine startup. Engine oil temperature, battery positive voltage (B+), and barometric pressure (BARO) are used by the ECM to calculate glow plug on time and the length of the duty cycle. On time normally varies between 10 and 120 seconds. With colder oil temperatures and lower barometric pressures, the plugs are on longer. If battery voltage is abnormally high, the duty cycle is shortened to extend plug life. The glow plug relay will only cycle on and off repeatedly when the system high voltage condition is greater than 16 volts. An open in the glow plug relay circuit will render the glow plugs inoperative. A short circuit could result in a glow plug continuous on condition. The glow plug light signal controls the WAIT TO START indicator light located on the instrument panel. When the light goes off, the engine is ready to be started. The light comes on every time a key-on reset occurs. On time normally varies between 1 and 10 seconds. The WAIT TO START light on time is independent of the glow plug relay on time because the glow plugs may stay on to improve performance until the engine reaches operating temperature. An open circuit in the glow plug light wiring will result in an inoperative glow plug light. A short circuit will result in a glow plug light continuous on condition.

An example of a typical engine that uses intake air heaters (IAH) to aid in starting the engine in cold conditions, and to improve control of white smoke is a Cummins 5.9 liter. Two separately operated heating grids may be operated together or separately by the ECM, based on input from the intake air temperature (IAT) sensor and the fuel temperature sensor. The four phases of operation are: Set-up, Preheat, Cranking, and Postheat. Set-up begins when the ignition is turned on and sensor inputs are updated. During preheat, both heaters are on and the "Wait To Start" lamp is illuminated, notifying the driver that the heaters are working and when to crank the engine. The duration of preheat is determined by intake air temperature. When the engine is cranking, both heaters are turned off. With the engine running, postheat begins and the ECM uses IAT and fuel temperature sensor input to activate the heaters based on postheat phase parameters.

Task E7   **Inspect, test, service, and replace EGR system components; including EGR valve, variable ratio/geometry turbocharger, cooler, piping, filter, electronic sensors, controls, system air pressure solenoids, and wiring.**

Due to stricter diesel engine emissions regulations imposed by the EPA, engine manufacturers are required not only to reduce particulate emissions, but are also required to reduce nitrides of oxygen or NOx emissions. NOx emissions are generally created when combustion chamber temperatures are above 1500°F. One method used to lower cylinder temperatures and reduce NOx emissions is to retard ignition timing. While this method will reduce cylinder temperatures, it will reduce engine performance and increase fuel consumption. Another way to reduce cylinder temperatures is to reduce the oxygen content in the cylinder, thereby lowering combustion temperatures. The diesel engine EGR system uses cooled exhaust gasses to "dilute" air charge's oxygen content to achieve lower temperatures. Normal ignition timing can be utilized and engine performance is usually not affected. The engine ECM controls the EGR system. The ECM monitors various engine parameters to determine when EGR gas should be allowed to flow—to provide power, economy, and pollution reduction. Several system designs are available to the market. The most common system uses EGR valve and controls, EGR cooler, variable geometry turbocharger (VGT), and electronic pressure and temperature sensors specific to the EGR system.

The EGR valve allows exhaust gas to be taken from the turbine side of the turbocharger, and routed to the cooler. The EGR valve may be operated by oil pressure, compressed air, or an electric motor. The VGT uses moveable vanes in the turbine outlet to vary the restriction of exhaust gas flow through the turbocharger, accomplishing two specific goals. First, it will help generate sufficient back pressure to enable exhaust gas flow; and second, it can be used to manipulate turbocharger speed, increasing the efficiency of the turbocharger. The VGT controls may also be operated by oil pressure, compressed air, or electric motors. The EGR cooler is simply a water-cooled heat exchanger, designed to cool the exhaust gas before it reenters the intake air. Various electronic sensors are used to monitor EGR gas temperature, gas flow, EGR valve position, and VGT vane position.

Task E8   **Inspect and repair exhaust brake system.**

Some transit vehicles utilize exhaust brakes to help slow the vehicle and increase brake lining life. Most exhaust brake systems are ECM controlled, using compressed air from the vehicle's air system to operate the actuator valve. The exhaust brake creates a restriction in the exhaust pipe after the outlet of the turbocharger. The valve consists of a moveable flap mounted in the exhaust pipe. When braking force is desired, an actuator rotates the flap blocking most of the exhaust flow exiting the engine. This, in turn, will place drag on the engine due to the resistance to airflow through the engine. The resistance is then transferred through the driveline to the wheels, slowing the vehicle. When braking assistance is no longer desired, the ECM will deenergize the solenoid, providing air pressure to the actuator. Spring pressure will force the flap to a neutral position not affecting exhaust flow. Although exhaust brakes are relatively inexpensive and simplistic in their operation, they are not very efficient, and can generate high exhaust temperatures when allowed to operate for extended periods of time.

F. Fuel System Diagnosis and Repair (13 Questions)

1. Mechanical Components (4 Questions)

Task F1.1 **Inspect, repair or replace fuel tanks, vents, cap(s), mounts, valves, screens, crossover system, supply and return lines, and fittings.**

In a diesel fuel subsystem, a clear divide exists between the suction side and the charge side represented by the fuel transfer pump. Most fuel subsystems are of this type and the terms suction circuit and charge circuit are used to describe each. A primary filter is most often located on the suction side of the transfer pump while the secondary filter is located on its charge side. However, in some fuel systems, notably those made by Cummins, all movement of fuel through the fuel subsystem is under suction. When such a fuel system uses multiple filters, the terms primary and secondary tend not to be used.

Fuel Tanks

Fuel is stored on commercial vehicles in fuel tanks. Many diesel fuel management systems are designed to pump much greater quantities of fuel through the system than that required for actually fueling the engine.

Generally, 1/2 to 2/3 of the fuel pumped is cycled through the fuel system and returned to the tank. The excess fuel is used to lubricate and cool high-pressure injection components, especially those exposed directly to the extreme temperatures of engine cylinders. As a cooling medium, the fuel transfers heat from the injection devices to the fuel tank, providing the fuel tank(s) with a role as heat exchanger.

Fuel tank vents should be routinely inspected for restrictions and protected from ice build-up. A plugged fuel tank vent will rapidly shut down an engine, creating a suction-side inlet restriction value the transfer pump will not be capable of overcoming.

To check for the presence of water in fuel tanks, first allow them to settle, then insert a probe (a clean aluminum welding rod) lightly coated with water detection paste through the fill neck until it bottoms in the base of the tank. Withdraw the rod and examine the water detection paste for a change in color. This test will give some idea of the quantity of water in the tank by indicating the height on the probe the color has changed to. Trace quantities (just the tip of the probe changes color) in fuel tanks are not unusual and will not necessarily present any problems.

Fuel Tank Sending Units

Most commercial fuel subsystems use remote (from the tank) fuel transfer pumps and not assemblies that incorporate the sending unit and a transfer pump. The fuel-sending unit is an integral assembly flange fitted to the tank, consisting of a float and arm connected to a variable resistor. As the float arm is moved through its stroke, the resistance changes, producing a reading at the dash gauge that correlates with fuel quantity. Fuel-sending unit problems can be diagnosed with a DMM (digital multimeter) in resistance mode, by moving the float arm through its arc and observing readings.

Pickup Tubes

Fuel pickup tubes are positioned so that they draw on fuel slightly above the base of the tank and thereby avoid picking up water and sediment. Pickup tubes are quite often welded into the tank; if they fail, the tank must be replaced. Fuel pickup tubes seldom fail but when they do it is usually by metal fatigue crack at the neck; this results in no fuel being drawn out of the tank by the transfer pump whenever the fuel level is below the crack.

Task F1.2 **Inspect, clean, test, repair or replace fuel transfer pump, lift pump, drives, screens, fuel/water separators/indicators, filters, heaters, coolers, ECM cooling plates, and mounting hardware.**

Fuel-Charging/Transfer Pumps

Fuel-charging or transfer pumps are positive displacement pumps driven directly or indirectly by the engine. A positive displacement pump displaces the same volume of fluid per cycle, increasing fuel quantity pumped proportionately with rotational speed. Similarly, if a positive displacement pump unloads to a defined flow area, pressure rise can be said to be proportional with rpm increase. On most truck and bus fuel systems, charging/transfer pumps are of the plunger or gear types.

Plunger-Type Pumps

Plunger-type pumps, which are often used with port-helix metering injection pumps, are usually flange mounted to the injection pump cam box and driven by a dedicated cam on the pump camshaft. Single-acting and double-acting plungers may be used, with the latter type specified in higher output engines requiring more fuel. A single-acting plunger pump (a pump with a single reciprocating element such as bicycle pump) has a single pump chamber and an inlet and outlet valve. Fuel is drawn into the pump chamber on the inboard stroke and pressurized on the outboard or cam stroke.

A double-acting pump has twin chambers each equipped with its own inlet and outlet valve. On the cam stroke, a two-way plunger charges the pump chamber while admitting fuel to the other. On the return stroke, the pump retraction spring reverses the process.

Gear-Type Pumps

Gear-type pumps, commonly used as transfer pumps, are normally driven from an engine accessory drive and are located wherever convenient. Gear pumps usually have an integral relief valve, which will define the peak-system charging pressure. Fuel injection systems designed to be charged at pressure values higher than typical, tend to use gear-type transfer pumps over cam-actuated, plunger pumps. In instances where a gear pump feeds an injection system with no main filter in series, a filter mesh is sometimes incorporated to protect the injection pumping apparatus. When gear pumps are used, there is a small chance that gear teeth cuttings can be discharged into the system. A majority of the full authority, electronic management fuel systems use gear-type pumps.

Hand-Primer Pumps

A hand-primer pump may be a permanent fixture to a fuel subsystem located on the fuel transfer pump body or a filter-mounting pad. A hand-primer pump can also be a useful addition to the technician's tool kit and it can be fitted to a fuel subsystem when priming is required. The function of a hand-primer pump is to prime the fuel system whenever prime is lost. Typically they consist of a hand-actuated plunger and use a single-acting pumping principle. On the outward stroke, the plunger exerts suction on the inlet side, drawing in a charge of fuel to the pump chamber; on the downward stroke, the inlet valve closes and fuel is discharged to the outlet. When using a hand-primer pump, it is important to purge air downstream from the pump on its charge side. Some fuel subsystems mount a hand primer to the transfer pump housing. Some newer fuel systems have self-contained, electric priming pumps whose function is to prime the system after servicing.

Fuel Filters

Fuel filters entrap particulate (fine sediment) in the diesel fuel and some current secondary filters will filter to the extent that water in its free state will not pass through the filtering media. A typical fuel subsystem with a suction circuit and a charge circuit will in most cases employ a two-filter arrangement, one in each of the suction and charge circuits. Two basic types of filter are used: the currently more common spin-on, disposable cartridge type and the canister and disposable element type. Spin-on filters are obviously easier to service and are the filter design of choice by most manufacturers.

Primary Filters

Primary filters represent the first filtration stage in a typical two-stage filtering fuel subsystem. Primary filters are usually under suction, plumbed in series between the fuel tank and the fuel transfer pump. They are designed to entrap particulate sized larger than $15–30\mu$ (microns) (1μ = 1 millionth of a meter) depending on the fuel system, and achieve this using media ranging from cotton-threaded fibers, synthetic fiber threads, and resin impregnated paper.

Secondary Filters

Secondary filters represent the second filtration stage in two-stage filtering. In a typical fuel subsystem, the transfer pump charges the secondary filter, enabling the use of more restrictive filtering media. The secondary filter would normally be located in series between the transfer or charging pump (the pump responsible for pulling fuel from the fuel tank and charging the fuel injection components) and the fuel injection apparatus. Current secondary filters may entrap up to 98 percent particulate sized as small as 1μ (1 millionth of a meter) but filtering efficiencies of $2–4\mu$ are more common.

Water in its free or emulsified states will not be pumped through many of the current generation of fuel filters. This results in the filter plugging on water and shutting down the engine by starving it for fuel. Secondary filters use a variety of media including chemically treated pleated papers and cotton filters.

In a fuel subsystem that is entirely under suction, such as Cummins PT, the terms primary and secondary are not used to describe multiple filters when fitted to the circuit. As every filtering device used in the fuel subsystem is under suction, the inlet restriction specification is critical and if it exceeded the maximum, it would result in a loss of power caused by fuel starvation.

Servicing Filters

Most fuel filters are routinely changed on preventative maintenance schedules that are governed by highway miles, engine hours, or calendar months. They are seldom tested to determine serviceability. When filters are tested, it is usually to determine if they are restricted (plugged) to the extent of reducing engine power by causing fuel starvation.

The transfer pump usually charges secondary filters. Testing charging pressure (the pressure downstream from the charging/transfer pump) is normally performed with an accurate, fluid-filled pressure gauge plumbed in series between the transfer pump and injection pump apparatus; it is not generally used as a method of determining the serviceability of a secondary filter. Secondary filters tend to be changed by preventative maintenance schedule rather than by testing or when they plug on water and shut down an engine.

Water Separators

Most current diesel engine-powered highway vehicles have a fuel subsystem with fairly sophisticated water removal devices. Water appears in diesel fuel in three forms: free state, emulsified, and absorbed. Water in its free state will appear in large globules and because of its greater weight than diesel fuel, will readily collect in puddles at the bottom of fuel tanks or storage containers.

Water emulsified in fuel appears in small droplets. These droplets are minutely sized, so they may be suspended for sometime in the fuel before gravity takes them to the bottom of the fuel tank. Absorbed water is usually water in solution with alcohol, a direct result of the methyl hydrate (type of alcohol added to fuel tanks as deicer or in fuel conditioner) added to fuel tanks to prevent winter freeze-up. Water that is absorbed in diesel fuel is in its most dangerous form because it can be pumped through the filters into the injection system.

Often water separators will combine a primary filter and water-separating mechanism into a single canister. They employ a variety of means to separate and remove water in free and emulsified states; they will not remove water from fuel in its absorbed state.

Use a mercury (Hg) manometer to find the water separator restriction. In cases where the entire fuel subsystem is under suction, the consequences of exceeding the restriction specification are generally more severe—the result being fuel starvation to the engine.

Servicing water separator units is a simple process but one that should be undertaken with a certain amount of care as it is easy to contaminate the fuel in the separator canister either by priming it with unfiltered fuel or by permitting dirt to enter when the canister lid is removed. Most water separators have a clear sump through which it is easy to observe the presence of water. All water separators are equipped with a drain valve to siphon water from the sump. Water should be routinely removed from the drain valve. The filter elements used in combination water separator/primary filter units should be replaced in most instances with the other engine and fuel filters at each full service. However, some manufacturers claim their filter elements have an in-service life that may exceed the oil change interval by two times. Whenever a water separator is fully drained, prime it before attempting to start the engine.

Fuel Heaters

In recent years, it is more common to find vehicles equipped with fuel heaters. In fuel systems where fuel is flowed through the injection system circuitry at a rate much higher than that required for fueling the engine, constant filtering of fuel removes some of the wax and therefore some of its lubricity even when the appropriate seasonal pour-point depressants are present.

An electric heating element uses battery current to heat fuel in the subsystem. This type of fuel heater offers a number of advantages, most notable is that the heater can be energized before startup so that cranking fuel is warmed up. Electric element fuel heaters may be thermostatically managed so that fuel is only heated as much as required and not to a point that compromises some of its lubricating properties.

Some fuel heaters use both electric heating elements and coolant medium heat exchangers and furthermore manage the fuel temperature. Optimally, fuel temperature should be managed not to exceed 90°F (32°C). Once fuel exceeds this temperature its lubricating properties start to diminish, resulting in reduced service life of fuel injection components.

Task F1.3   Check fuel system for air and temperature; determine needed repairs; prime and bleed fuel system; check, repair or replace primer pump.

To troubleshoot the source of air admission to the fuel subsystem, a diagnostic sight glass can be used. The fuel subsystem consists of a clear section of tubing with hydraulic hose couplers at either end and is fitted in series with the fuel flow. However, the process of uncoupling the fuel hoses will always admit some air into the fuel subsystem, so the engine should be run for a while before reading the sight glass.

Priming a Fuel System

Most OEMs prefer that the technician avoid pressurizing air tanks to prime a fuel system. Remember, diesel fuel contains volatile fractions and the act of pressurizing a fuel tank with air pressure will vaporize some fuel while air exiting an air nozzle creates friction and the potential for ignition. Avoid this practice in extreme hot weather conditions.

Recommended Priming Procedure

When a vehicle runs out of fuel and it is determined that the fuel subsystem requires priming, remove the filters and fill with filtered fuel:

- Locate a bleed point in the system. Often on an in-line injection pump system, the bleed point will be at the exit of the charging gallery. Crack open the coupling.

- Next, if the system is equipped with a hand-primer pump, actuate it until air bubbles cease to exit from the cracked open coupling. If the system is not equipped with a hand-primer pump, fit one upstream from the secondary filter and actuate until air bubbles cease to exit from the cracked open coupling.

- Retorque coupling. Crank the engine for 30-second segments with at least 2-minute intervals between cranking until it starts. This will allow for starter motor cooldown. In most diesel engine systems, the high-pressure circuits will self-prime once the subsystem is primed.

Task F1.4   Inspect, test, repair or replace high/low pressure systems (check valves, pressure regulator valves and restrictive fittings).

The fuel system must be capable of supplying enough pressurized fuel to the injection pump or injectors to meet engine demands. One key to meeting this demand is to ensure that the fuel pressure remains constant during RPM variations. If the fuel pressure can not be stabilized, the injection pump or injectors may experience fuel starvation and a loss of engine performance can result. Also, ensure protection against overpressurization, as this can cause leaks and possible engine damage. Usually, most means of pressure control involve calibrated spring pressure acting against a valve. When pressures are below maximum set points, the spring tension holds the valve closed, allowing pressure to achieve desired levels. When pressures rise, the spring will collapse, allowing the valve to move in its bore, bleeding the excess pressure off, usually back to the pump inlet. Problems with the spring or the valve sticking in the bore can cause unregulated pressure.

Task F1.5   Inspect, adjust, repair or replace mechanical engine throttle and controls.

Throttle linkage adjustments are critical on older mechanical engines. The travel of the throttle controls the engine over its complete operating range. Check the travel from idle to the full fuel position; this must meet manufacturer's calibration specifications. If it is incorrect, adjustment of the stop screws will correct the problem. Before making any critical adjustments, ensure proper fuel pressure and flow is available. Check fuel filters for restrictions, sticking injector and throttle linkages, and low fuel pressure from worn pumps.

Task F1.6   Perform on-engine inspections, tests, adjustments, and time, or replace and time, distributor-type injection pumps.

The distributor-type pump and injection nozzle system is distinctive in appearance because the pump looks like a distributor. In fact, it operates much like an electrical distributor in that it distributes fuel to each cylinder at the appropriate time. An internal valve is rotated to line up with a high-pressure line dedicated to only one of the injectors. When the valve and line are aligned, fuel flows to that injector and it fuels the cylinder. Further rotation aligns the valve with the next injector's line and the next injector fires. This process is repeated in firing sequence. Fuel is metered and pressurized in the injector pump. The injectors atomize the fuel for combustion. To cool the injector tips, a portion of the fuel is returned to the pump by a nozzle leak-off line.

Task F1.7   Perform on-engine inspections, tests, and adjustments, or replace mechanical unit injectors.

In a mechanical unit injector fueled engine, each engine cylinder has its own unit injector, essentially a rocker actuated, pumping and metering element, and hydraulic injector nozzle combined in one unit. Each mechanical unit injector (MUI) has fuel delivered to it at charging pressure, variable between 30–70 psi (200 kPa–470 kPa) dependent on engine speed. The MUIs are responsible for metering, pumping to injection pressure values, and atomizing fuel directly to the engine cylinders. In the MUI, the pump is actuated by a dedicated cam profile on the engine camshaft. The high-pressure pipe is eliminated, and the hydraulic injector nozzle is integral with the assembly. The MUI is installed within a sleeve bored to the cylinder head, directly exposed to engine coolant passages. The injector cam profile actuates a train consisting of a roller-type lifter, pushrod, and rocker arm. The rocker arm acts on the tappet assembly of the MUI using a floating button as an intermediary to prevent side loading of the tappet by the rocker arm. The plunger is linked to the tappet and reciprocates within a stationary barrel. It is milled with a helix that can be classified as a lower helix design, resulting in a constant beginning, variable ending of fuel delivery timing. With this design of helix, the beginning of the effective pumping stroke is determined by the angular location of the cam. The barrel is machined with an upper port and a lower port diametrically offset. Metering of fuel is accomplished by rotating the plunger in the barrel bore and altering the point of register of the spill ports with the helix. The plunger is rotated by means of a gear, which is tooth-meshed to the rack. The gear collars the plunger in a manner that permits plunger rotation while it reciprocates.

Charging pressure is generated by a positive displacement gear pump, and responsible for all movement of fuel through the fuel subsystem. Engine output is controlled by a mechanical governor (in most truck and bus applications) that regulates fuel quantity by controlling the unit injector fuel racks.

Jumper pipes connect the fuel manifolds in the head with the unit injectors; one charges the unit injector, the other returns fuel to the return manifold. Fuel is cycled through the system whenever the engine is running; that is, all of the fuel delivered to the MUI is not used for fueling the engine. Fuel cycled through the MUI circuitry is used for cooling and lubricating the internal components of the assembly.

Set Injector Racks

Setting the injector racks is a critical adjustment that must be precisely executed to balance the fueling to each engine cylinder. Erratic setting of the MUI racks during tune-up is a main cause of hunting, a rhythmic fluctuation in engine rpm. The exact linear position of each MUI rack is determined by its control lever adjustment. This setting procedure varies by engine series and by manufacturer but must be precisely observed.

Task F1.8   Inspect, test, repair or replace fuel injection nozzles.

Hydraulic injectors, normally classified by nozzle design, are simple hydraulic switch mechanisms whose functions are to atomize and inject fuel into the engine cylinders. Of the three basic types of injector nozzles, two are effectively obsolete in medium and heavy-duty truck and bus engine applications. Poppet and pintle nozzles are best suited to indirect injected engine applications and therefore are not used currently in North American medium and heavy-duty engines. Multiorifice nozzles are used in most DI (direct injected) diesel engines using in-line, port-helix metering, injection pumps, and as an integral subcomponent of mechanically and electronically controlled, unit injection systems.

All current, high-speed diesel engines found in highway applications use hydraulic nozzles with the exception of the open nozzle designs used in Cummins hydromechanical and electronic common rail systems.

An injector nozzle is a hydraulic switch. One of its primary functions is to define the pressure required to trigger its opening. The term popping pressure is also used to describe the opening pressure of a nozzle. The actual nozzle opening pressure (NOP) value is defined by the mechanical spring tension of the injector spring. This spring tension loads the nozzle valve onto its seat and determines the hydraulic pressure required to unseat the valve. Most injectors incorporate a means of adjusting the injector spring tension so that the NOP value can be set to specification. The spring-tension adjustment mechanism is either shims or an adjusting screw and locknut. The NOP value is always one of the first performance specifications to be evaluated when testing injector nozzles on a bench test fixture (pop tester).

Removal of Injectors from the Cylinder Head

The OEM-recommended method of removing an injector from a cylinder head must be followed; failure to do so can result in failed injectors. Before attempting to remove injectors, thoroughly clean the surrounding area, remembering that any dirt around the injector bore can end up in the engine cylinder below it. Thin-bodied, pencil-type nozzles are vulnerable to any side-load force and the correct pullers should always be used. Many injector assemblies are flanged in which case, the hold down fasteners and clamps should first be removed and the injector levered out using an injector heel bar. The injector heel bar is 8–12 in. (20–30 cm) in length to prevent the application of excessive force to the injector flange. Where cylindrical injectors are used, a slide hammer and puller nut that fits to the high-pressure inlet of the injector should be used to pull the injector. With certain cylindrical hydraulic injectors, the high-pressure delivery pipe fits to a recess in the injector through the cylinder head; the high-pressure pipe must be backed away from the cylinder head before attempting to remove the injector or both the injector and the seating nipple on the high-pressure pipe will be damaged. When removing injectors from cylinder heads, remove the injector nozzle washer at the same time. Do not reuse injector washers. Both steel and copper washers harden in service.

When the injector has been removed, use plastic caps to seal the cylinder head injector bores, the injector inlet, and high-pressure pipe nipples. Ensure that a set of injectors is marked by cylinder number and properly protected in an injector tray (wrapping the injectors in shop rags will do if no injector tray is available).

Testing

Consult OEM specifications and procedures before the nozzle to be tested is placed in the nozzle test fixture. A typical procedure for testing nozzle assemblies is:

- Clean injector externally with a brass wire brush.

- Locate manufacturer's test specifications.

- Mount the injector in the bench test fixture. Build pressure slowly using the pump arm, watching for external leakage.

- Bench test NOP value and record. Use three discharge pulses and record the average value. Sticking or a variation in NOP value that exceeds 10 atms (150 psi) fails the nozzle.

- Test forward leakage by charging to 10 atms (150 psi) below the NOP value and holding the gauge pressure at that value while observing the nozzle. Any leakage evident at the tip orifice fails the nozzle.

- Check the back leakage factor by observing pressure drop from a value 10 atms (150 psi) below NOP. Pressure-drop values should typically be in the range of 50–70 atms (700–1000 psi) over a 10-second test period. Pressure drop that exceeds the OEM specification indicates too little valve-to-body clearance (possibly caused by valve-to-body mismatch). A rapid pressure drop exceeding the OEM specification indicates excessive nozzle valve to body clearance, a condition usually caused by wear.

- Again, actuate the bench fixture pump arm and observe the nozzle spray pattern, checking for orifice irregularity. Ignore nozzle chatter (rapid pulsing of the nozzle valve): this can be regarded as a test bench phenomenon due to slow rate of pressure rise. In some modern injectors, the intensity of chatter noise is reduced due to the use of double seats.

Reinstallation of Injectors

Clean the injector bore before installation using a bore reamer if necessary to remove carbon deposits. Blow out the injector bore using an air nozzle. Most diesel engine OEMs prefer that injectors be installed dry: lubricants such as Never-Seize can hinder the ability of the injector to transfer heat to the cylinder head. Turn the engine over and pump fuel through the high-pressure pipe prior to connecting them to the injector. This procedure will purge the pipe of air and possibly debris that may have intruded into the line while it was disassembled. Always torque the nut on the high-pressure pipe with a line socket. Failure to this can result in damage to both the line nut and the nipple and seat.

Task F1.9   Inspect, adjust, repair or replace smoke limiters (air/fuel ratio controls).

An aneroid is a low-pressure sensing device. In application, it is used on a turbocharged diesel engine to measure manifold boost and limit fueling until the boost pressure achieves a predetermined value.

Such devices are known as puff limiters, turbo boost sensors, AFC valves, and smoke limiters, and all seek to accomplish the same objective. They typically consist of a manifold within which is a diaphragm; boost air is piped from the intake manifold to act on the diaphragm. Such devices are used on most current boosted engines (turbocharged engines).

When an aneroid is used on an in-line, port-helix metering pump, it is usually a mechanism consisting of a manifold, spring, and control rod. The manifold is fitted with a port and a steel line connects it directly to the engine intake manifold. In this way, boost pressure is delivered to the aneroid manifold where it will act directly on the diaphragm within. Attached to the diaphragm is a linkage connected either directly or indirectly to the fuel control mechanism (rack). A spring loads the diaphragm to a closed position that will limit fueling by preventing the rack from moving into the full fuel position. When manifold boost acting on the diaphragm is sufficient to overcome the spring pressure, it acts on the linkage, permitting the rack full travel and thus maximum fueling. Such systems are easily and commonly shorted out by operators in the mistaken belief that aneroid systems reduce engine power. The emission of a "puff" of smoke from the exhaust stack at each gear shift point is an indication that someone has tampered with the aneroid/boost sensing mechanism.

An altitude compensator contains a barometric capsule that measures barometric pressure. Based on this pressure, it down rates engine power at higher altitudes to prevent overfueling. They are required when running at higher altitudes because the oxygen density in the air charge decreases with an increase in altitude and unless the fuel system is aware of this, it effectively overfuels the engine. The critical altitude at which some measure of injected fuel quantity derate is required to accord with today's low emission engines is 1000 ft. (303 meters) in altitudes. Older engines with altitude compensators usually did not derate the fuel system until an altitude of at least 2000 ft. (606 meters).

Current engines use variable capacitance and Piezio resistive sensors to signal manifold boost and atmospheric pressure conditions to the ECM. The ECM therefore modulates engine fueling based on these signals.

Task F1.10   Inspect, reinstall or replace high-pressure injection lines, fittings, seals, and mounting hardware.

To remove high-pressure fuel lines:

• Remove the low-pressure or leak-off fuel lines if used.

• Disconnect the fuel injection lines at the injectors. Cap each injection line and injector.

• Disconnect the fuel lines at the fuel injection pump. Cap each line and pumping element as the line is removed to prevent the entry of dirt in the system.

Once removed from the engine, individual fuel lines can be disconnected from the assembly for replacement. Keep in mind when ordering parts that each line is different and cannot be interchanged; therefore, order lines individually for the particular cylinder of that the line.

To install high-pressure fuel lines:

• Make sure that the clamps holding the lines together are tight. High-pressure fuel lines must be installed as a set. It is much easier than installing them individually.

• Wiggle the high-pressure fuel injection lines into position on the engine. Start from the back and go toward the front of the engine.

• Install the injection lines onto the injection pump and tighten the nuts. Tighten the brackets holding the high-pressure fuel lines to the intake manifold.

• Connect the high-pressure fuel injection lines to the fuel injectors one at a time, starting with cylinder #1. Do not remove the caps from the fuel lines or components until they are to be connected; this helps prevent the entry of dirt into the system.

• Tighten the fuel injection line to the injector nuts and torque to specification. Do not overtighten as this may collapse the sealing nipple, reducing flow to the nozzle.

• Install the low-pressure fuel lines and filter assembly back on the engine. Tighten the nuts to specifications.

• Run the engine and check for leaks.

Note: Reverse procedure for installation.

Task F1.11   Inspect, test, adjust, repair or replace engine fuel shut-off devices and controls, including engine protection shut-down devices, circuits and sensors.

A typical fuel shut-off solenoid is comprised of two coils within one housing. A powerful "pull" coil is used to pull the fuel shut-off lever to the "open" position. The pull coil is fed from the switched side of the starter motor relay, so it is energized whenever the ignition switch is in the Start position. At the same time, a less powerful "hold" coil is fed from a "Start/Run" circuit. When the ignition switch is released to the Run position, the pull coil drops out and the hold coil is left to keep the fuel shut-off lever in the "open" position. Because pull coils draw a considerable amount of current while energized, do not allow them to remain energized for longer than 30 seconds. In instances of a no-start condition in which the engine turns over normally and all correct starting procedures are being followed, look at the fuel shut-off solenoid to make sure it is being actuated. In some cases, it may be necessary to adjust the travel of the arm on the fuel solenoid. To adjust the fuel shut-off solenoid check the manufacturer's procedure.

2. Electronic Components (9 Questions)

Task F2.1 Check and record engine electronic diagnostic codes and trip/ operational data; clear codes; determine needed repairs.

Modern engines are controlled and monitored by electronic control modules (computers), which are capable of storing fault codes that identify circuit or component malfunctions. To retrieve these codes, the technician connects a diagnostic reader to a data link. Some vehicles will display a limited readout of fault codes by blinking an instrument panel indicator in a two-digit code called a flash code.

Electronic service tools capable of reading ECM data are connected to the on-board electronics by means of a SAE (Society of Automotive Engineers)/ATA (American Trucking Association) J1587/ J1708/1939 6- or 9-pin Deutch connector in all current systems. This data connector is used by all the engine electronics OEMs and is most often referred to as an ATA connector. This common connector and the adherence by the engine electronics OEMs to SAE software protocols enables proprietary software of one manufacturer to at least read the parameters and conditions of their competitors. Therefore, if a Navistar-powered vehicle has an electronic failure in a location where the only service dealer is DDC, some basic problem diagnosis can be undertaken using the DDC electronic diagnostic equipment.

Most current engine management systems will do much more than simply log codes. They communicate with a chassis management computer to which all the other onboard electronic circuits are connected and optimize the running of the engine as the conditions vary. While electronic system management modules are obliged to produce the universal fault codes that can be at least read by their competitors, each usually incorporates self-diagnostics that greatly exceed the minimum requirements. In addition, the computers that are used to manage engines can also track trip and operational data that can be downloaded to off-vehicle computer systems for analysis. Control modules are designed to monitor all the systems and components they manage. The self-diagnostic programming in the module enables the system to identify a circuit or component that is not functioning properly. When this happens a fault code is produced. There are two classifications of fault codes: active and inactive (historic). Active codes are faults that exist at the time of the self-diagnostic scan. An active code cannot be "erased" from the system until it is corrected. When the cause of an active fault code is identified and repaired, the code becomes inactive but continues to be displayed until erased by the technician. An intermittent problem, such as a sensor that produces erratic readings, could also produce an inactive code if the problem sensor was producing valid input at the time of the diagnostic scan. The inactive code would enable the troubleshooting technician to identify the erratic circuit or component even if it is not producing out of specification at the time of the diagnostic scan.

Removing inactive codes must be performed using the manufacturer's procedure. Removing inactive codes in a heavy-duty engine management system is usually performed using a reader/programmer such as Pro-Link or a PC equipped with system-specific software and connected to the vehicle data link (ATA connector).

Accessing information from a heavy-duty electronic control module may be accomplished using each manufacturer's specific electronic service tools (ESTs), by using PC-based software that is available by license agreements from the manufacturers, or by using a generic EST such as the Pro-Link 9000, which has become an industry standard. Each manufacturer's ESTs and PC-based software by agreement will access and retrieve information from other manufacturers' products, although some information is regarded as proprietary and so each usually features self-diagnostics that greatly exceed the minimum requirements. Retrieved fault codes will be universal codes that are preceded by identifiers. There are four groups of identifiers; message identifiers (MIDs), parameter identifiers (PIDs), failure mode identifiers (FMIs), and subsystem identifiers (SIDs). MID refers to the major vehicle system that is the source of the data transmission, such as the engine, transmission, or brakes.

The MID is the first character of every J 1587 message. PID refers to electronic components within an electronic subsystem. It is a 2- or 3-digit number that identifies the component, and may be followed by a single digit that describes the type of failure (failure mode identifier, or FMI). SID is a 1- or 2-digit number assigned to identify a section of a control system that does not have a related PID. It identifies major subsystems which may be repaired or replaced and for which failures may be detected, such as the cruise control switches, clutch switch, and data links. When using Pro-Link, make sure that you have the correct cartridge installed. A multiprotocol cartridge is available, which accepts a variety of software cartridges for engine, transmission, and brake systems. You will need the correct adapters and multipin connectors to access the system. Most vehicles are equipped with either a 6- or 9-pin Deutch SAE/ATA J1587/J1708/J1939 connector, which is usually just referred to as the ATA connector. The connector is located in the engine compartment, or in the dash, usually to the left of the steering column. This common connector and the software protocols established by agreement enable proprietary software used by various OEMs to communicate and at least read the parameters and conditions of their competitors. Most engine management systems must also communicate with other chassis systems to optimize engine performance. Begin by ensuring that the ignition switch and the diagnostic equipment are off; then connect your diagnostic equipment and turn it on. Turn on the ignition and observe whether communication is established. You will be presented with a series of prompts or menu choices that access different portions of the software. When codes are present, it is usually useful to record the codes and then clear them and retest to see whether they reappear. Codes that reappear indicate an active code that must be repaired before proceeding. Use a digital multimeter to verify sensor values and compare to specifications. Refer to manufacturer's manuals for diagnostic flow charts and symptom-based strategies.

Task F2.2   Inspect, adjust, repair or replace electronic throttle and PTO (high/low idle) control devices, circuits, and sensors.

The throttle position sensor (TPS), a variable resistor that signals the ECM the accelerator pedal angle, is supplied with a 5-volt reference and returns as a signal a portion of the reference voltage. The TPS has the highest resistance at 0 pedal angle. At full pedal travel, the resistance is lowest and the signal to the ECM will be near 5 volts. Some engine manufacturers also use an Idle Validation Switch (IVS) in conjunction with the TPS as a confirmation to the ECM as to the accelerator pedal's position. The ECM can compare TPS voltage readings against the IVS position (on-idle or off-idle position). If the ECM detects incorrect TPS voltage when IVS dictates an idle position, the ECM will ignore the incorrect TPS signal, allowing the engine to idle only. This can prevent the engine from accelerating due to wiring or sensor faults.

Usually operation of a typical TPS can be bench tested or connected with the circuit. Bench testing may be performed with the sensor isolated from the circuit using a digital multimeter (DMM) set to read resistance. This method is not recommended as it is not accurate enough to identify flat spots and shorts/opens that occur outside the sensor. The best method to verify TPS performance is to use the system self-diagnostics. This is performed by connecting a reader/programmer or PC with system specific software. Most current systems will guide the technician through the test. The onboard diagnostics will test the sensor dynamically by analyzing the return signal voltage through the pedal stroke and is therefore accurate.

Task F2.3   Perform on-engine inspections, tests, and adjustments on hydraulic electronic unit injectors (HEUI) and electronic controls (rail pressure control).

A disadvantage of the electronic unit injector (EUI) systems is that the fueling window is defined by the hard parameters of the injector train cam profile. The Cat/Navistar HEUI (hydraulically actuated, electronic unit injector) system uses engine lube oil as hydraulic medium to actuate the fuel delivery pulse in their HEUI assemblies. The delivery stroke is actuated hydraulically, switched by the engine management ECM(s), and now not confined to any hard limits. HEUI is therefore truly a full-authority system in that the ECM can select the fuel pulse (pump effective stroke) to occur at anytime it calculates that it is required. Perhaps more importantly, HEUI represents a critical step toward eliminating the camshaft from a production diesel engine and achieving a fully regulated engine environment.

HEUI management can be loosely divided into subsystems as described below.

Fuel Supply System

The fuel supply system delivers fuel from the vehicle's tanks to the injector units. Fuel movement through the fuel supply system is the responsibility of a cam driven plunger pump flange mounted to the engine block rear of the high-pressure oil pump. This transfer pump pulls fuel from the chassis fuel tank(s) through a fuel strainer. It then charges fuel through a disposable cartridge-type fuel filter and feeds it to the fuel gallery of fuel/oil supply manifold. A fuel pressure regulator at the fuel manifold outlet is responsible for maintaining a charging pressure of approximately 30 to 60 psi. Fuel is cycled through the fuel supply system and the HEUIs are mounted in parallel from the fuel manifold.

Injection Actuation System

The HEUI system uses hydraulically actuated, electronically controlled injector assemblies to deliver fuel to the engine's cylinders. The hydraulic medium used to actuate the pumping action required of the injector is engine oil. The engine lubrication circuit provides a continuous supply of engine lube to the HEUI high-pressure pump, a gear-driven, swash plate hydraulic pump used to boost up the lube oil pressure to values exceeding 3000 psi.

Actual high-pressure oil values are managed by the IPR or injection pressure regulator, which is controlled electronically and actuated electrically. Swash plate pumps use opposing cylinders and double-acting pistons driven by a swash plate. They are similar in principle of operation to common automotive A/C compressors. The injection pressure regulator manages the high-pressure oil pressure values between a low of 485 psi to highs exceeding 3500 psi by receiving all the lube pressurized by the high-pressure pump and spilling the excess to the oil reservoir.

The high-pressure oil is then piped to the oil gallery ducting within the fuel/oil supply manifold. From there, the oil is delivered to an exterior annulus in the upper portion of each HEUI. The HEUIs are mounted in parallel and fed by the high-pressure oil manifold. When the HEUI solenoid is energized, a poppet valve is opened by an electric solenoid within the HEUI. This permits the high-pressure lube to flow into a chamber and act on the amplifier piston (Navistar)/intensifier piston (Caterpillar), actuating the pumping stroke required to convert the HEUI charging pressure to injection pressure values. At the completion of the duty cycle or PW (pulse width), the HEUI solenoid cartridge is deenergized and the poppet valve retracts, spilling the oil acting on the amplifier piston to the rocker housing.

HEUI (Hydraulic Electronically Controlled Unit Injector)

The HEUI is an integral pumping, metering, and atomizing unit controlled by ECM switching apparatus. The unit is essentially an EUI that is actuated hydraulically rather than by cam profile. At the base of the HEUI is a hydraulically actuated, multiorifice nozzle. When the HEUI pumping element achieves the required NOP (nozzle opening pressure) value acting on the sectional area of the nozzle valve exposed to the pressurizing annulus, the valve retracts, permitting fuel to pass around the nozzle seat and exit the nozzle orifice directly to the engine cylinder. A VCO (valve closes orifice) nozzle design is used.

The amplifier or intensifier piston is responsible for creating injection pressure values. (This component, termed an amplifier piston in Navistar technical literature and an intensifier piston in Caterpillar versions, is referred to as an amplifier piston in this text.) When the HEUI is energized, high-pressure oil supplied by a stepper pump acts on the amplifier piston and drives its integral plunger downward into the fuel in the pumping chamber.

A passage connects the pump chamber with the pressure chamber of the injector nozzle valve. The moment the HEUI is deenergized, the oil pressure acting on the amplifier piston is bled off and the amplifier piston return spring, plus the high-pressure fuel in the pump chamber, retracts the amplifier piston/plunger, causing the almost immediate release of the pressure holding the nozzle valve open. This results in rapid cessation of the injection pulse.

HEUIs typically have NOPs of 5000 psi with a potential for peak pressures of up to 24,000 psi. As a general rule, the oil pressure acting on the HEUI amplifier piston is "amplified" by a factor of 7:1 in the fuel pump chamber. The droplet sizing is designed to decrease as the length of the injection pulse increases (as the real-time dimension available for combustion decreases); the rapid pressure collapse enabled by HEUIs avoids the injection of larger sized droplets toward the end of injection that would be difficult to burn completely. At the completion of the HEUI duty cycle or PW, the pressurized oil that actuated the pumping action is spilled to the rocker housing.

Plunger descent rate is fully managed by the ECM, which switches and controls the actuating high-pressure oil value through the IPR (injection pressure regulator). HEUI injectors are capable of being driven or switched at high rates and the latest versions have plunger and barrel geometry that provides automatic pilot injection. The term pilot injection is used to describe an injection pulse that is broken into two separate phases. In a pilot injection fueling pulse, the initial phase injects a short-duration pulse of fuel into the engine cylinder, ceases until the moment of ignition, and at that point, resumes injection, pumping the remainder of the fuel pulse into the engine cylinder. Pilot injection has been used as a cold start and warmup strategy in EUI systems to avoid excess fuel in the engine cylinder at the point of ignition and as a result, minimize the tendency to cold start detonation. HEUI systems with the pilot injection feature produce a pilot pulse for each injection. The input circuit of the HEUI system is comparable with that of any other full-authority system permitting engine management to optimize operation for changing conditions, monitor faults, provide engine protection, and meet emission requirements. HEUI engine management is found on Caterpillar 3126, C-7, C-9, and Navistar 444, 466, and 530 engines.

Task F2.4   Perform on-engine inspections, tests, and adjustments on electronic unit injectors (EUI) and electronic controls.

Currently, most buses use full-authority, electronic unit injector (EUI) systems. The fuel subsystem used with EUI fueling consists of a suction and charge circuit with fuel movement provided by a positive displacement gear-type pump. Fuel from the fuel subsystem is provided to the EUI by means of a manifold integral with the cylinder head. Fuel is routed through the EUI for purposes of fueling the engine with excess fuel used to lubricate and cool.

The EUI is a cam-actuated, integral pumping, metering, and atomizing device that is controlled by the vehicle management computer or ECM. A typical EUI can be divided into circuits as described below.

Control Cartridge

The control cartridge is an ECM-switched solenoid with a poppet valve integral with its armature. The poppet valve is held open by spring force. When the poppet valve is open, fuel charged to the EUI by the fuel subsystem is routed through the EUI internal circuitry and allowed to escape to spill into the return circuit. When the ECM energizes the control solenoid, the armature is sucked into the coil, closing the return passage in the EUI and preventing escape of fuel to the return circuit. The ECM energizes the control solenoid by means of a pulse width modulated signal. The energized cycle of the solenoid is known as duty cycle or pulse width (PW). Whenever the solenoid cartridge is energized, fuel is trapped in the EUI pump chamber and the EUI is pumping fuel.

Plunger and Barrel

The pumping element of an EUI consists of a plunger and barrel. The barrel is the stationary member of the pumping element; the plunger, the reciprocating member, is spring loaded into its retracted position. The plunger is actuated by cam profile. The injector camshaft may be cylinder block mounted or overhead. When the injector train is actuated by cam profile, the EUI plunger is driven through its stroke. The actual plunger stroke does not vary regardless of how the engine is being fueled. Some electronic unit injectors require setting of the EUI plunger height, a mechanical adjustment performed by adjusting an adjustable screw on the rocker arm. Since plunger height affects timing, this should be accurately adjusted.

Fuel from the fuel subsystem is routed into the EUI pump chamber and passes on to the circuitry in the control cartridge. When the control cartridge solenoid is not energized, fuel is simply allowed to flow through this circuit. When the plunger is actuated by the injector cam, fuel in the pump chamber is displaced. However, an effective stroke cannot begin until the control cartridge is energized. This traps fuel in the EUI circuit, specifically in the pump chamber. As the plunger is driven downward through its stroke, fuel is pressurized to injection pressure values.

Nozzle Assembly

Connected to the EUI pump chamber by means of a duct is a nozzle assembly. This is a simple multiorifice nozzle assembly, a hydraulic switch set to open at a preset hydraulic pressure. The nozzle assembly functions to define nozzle opening pressure (NOP) and atomize the fuel charge to the cylinder. Typical NOPs for EUIs are around 5000 psi.

EUI systems are full-authority electronic engine management systems, with comprehensive programming, self-diagnostic, and data management capabilities. Input circuit components are similar to other late-generation systems. Typical EUI systems are found on Caterpillar C-10, C-12, C-13, C-15, and C-16 engines; Cummins CELECT; Detroit Diesel Series 50 and 60; and Volvo VECTRO managed engines.

Injector Response Time and Pulse Width

Injector response time (IRT) and pulse width (PW) are indicators of the overall electrical integrity and power balance of the engine. Both parameters can be displayed on any electronic service tool (EST). IRT is the measurement of the time elapsed between the ECM energizing the solenoid of the injector until the poppet valve actually closes. IRT is measured in milliseconds. If the response time is too long or too short compared to specifications, it normally indicates a circuit problem with the control coil. PW is the amount of time the control coil is energized measured in degrees of crankshaft rotation. The longer the coil is energized, the greater degrees of rotation will occur. This directly relates to more fuel being injected.

Injector Cut-out Test

There is no appropriate way mechanically to short out a EUI, so when a cylinder misfire is being diagnosed, the ECM electronics perform this task by electronically cutting out the EUI in sequence and analyzing the performance effect. When a ECM commands an engine to run a specific rpm, and an injector cut-out test is performed, first the average duty cycle of the EUI in degrees of pulse width (PW) is displayed. If the engine rpm is to be maintained during the test (this test is normally run at idle or 1000 rpm), as each EUI is cut out electronically by the ECM, the PW of the remaining injectors will have to be lengthened if engine rpm is held at the test value. This would be true until a defective EUI was cut out; in this case, there would be no increase in the average PW of the operating EUI. If a 6-cylinder engine had 1 dead EUI the following would occur if an injector cut-out test sequence were performed with the engine running at 1000 rpm. To run the engine at 1000 rpm, the 5 functioning EUIs would produce fueling values read in PW degrees, which would be averaged on the display. When 1 of the 5 functioning EUI is electronically cut out by the ECM, only 4 EUI would be available to run the engine at the test speed of 1000 rpm, causing the average PW to increase. As each functioning EUI was cut out in sequence, the average PW would have to increase to maintain the test rpm. However, when the defective EUI was cut out, there would be no change in the average PW because it was based on the engine running on 5 cylinders.

Task F2.5 Inspect, test, adjust, repair or replace engine electronic fuel shut-down components, circuits, and sensors, including engine protection and automatic stop systems.

Most full-authority engine management systems monitor a range of vehicle and chassis functions and are programmed with engine protection strategy that may derate or shut down the engine when a running condition could result in catastrophic failure. The operator is alerted to system problems by illuminating dash lights, the check engine light (CEL), and the stop engine light (SEL). The CEL is illuminated when a system fault is logged to alert the operator that system fault codes have been generated. The SEL is illuminated when the ECM detects a problem that could result in a more serious failure. The system may also be programmed to shut down or ramp down to idle speed after such a problem has been logged. A shutdown override switch, known as an SEO or stop engine override switch, will permit a temporary override of the shutdown command.

Engine protection strategy can be programmed to 3 levels of protection beginning at a low level of protection, essentially a driver alert, and culminating with a complete shutdown of the engine. The level of protection is selected by the vehicle owner and programmed to the vehicle electronics as a customer option. As a customer option, the failure strategy can be changed at any time by a technician equipped with the access password and an EST with the appropriate software.

Parameters monitored:

- Low coolant pressure
- High coolant temperature
- Low oil pressure
- Low coolant level
- High oil temperature
- Auxiliary digital input

When the engine management system detects a fault that may result in catastrophic engine damage, it can be programmed to react to the problem in 3 ways:

1. Warning only. The CEL and/or SEL will illuminate but it up to the operator to take action to avoid potential engine damage. No power or speed reduction will occur. Engine protection is left entirely to the discretion of the operator.

2. Ramp down. The CEL and/or SEL will illuminate when a fault is detected and the ECM will reduce fuel quantity injected over a 30-second period after the SEL illuminates. The fuel quantity maintained for initial ramp down is that operative just prior to the logging of the fault condition. A diagnostic request/stop engine override switch is required when this engine protection option is selected.

3. Shutdown. Both the CEL and SEL will illuminate and the engine will shut down 30 to 60 seconds afterwards. Once again, a diagnostic request/stop engine override switch is required if this option is selected.

Stop Engine Override Options
Two types of stop engine override are available:

1. Momentary override. This stop engine override—resets the 30- to 60-second shutdown timer, restoring engine output to the value operative at the moment the SEL was illuminated. To obtain subsequent overrides, the switch must be recycled after 5 seconds. Most systems will record the number of overrides actuated after the triggering fault occurrence.

2. Continuous override—option 1. This stop engine override option is used when a vehicle needs full power during a shutdown sequence. Full power capability is maintained for as long as the override switch is pressed. It is intended for coach applications only.

Task F2.6   Inspect and test voltage, ignition, and ground circuits and connections for electrical/electronic components; determine needed repairs.

With modern electronic modules and low-level signals, grounding between modules has become extremely important. Even a resistance of 1 to 2 ohms between ground can cause a fault to occur. Test grounds without power applied using a digital ohmmeter between the ground lead and a clean common ground. Do not use a standard ohmmeter as the internal battery could cause damage to the module. Test grounds with power applied using a digital voltmeter between the ground lead and a clean common ground. Do not use an analog voltmeter because its internal resistance can affect the internal circuit of the module. Once a defective ground has been located, clean the ground connection and treat it with conductive grease. Fasten all leads securely and connect a suitable wire between common grounds. Retest all circuits to ensure proper operation.

Task F2.7   Inspect and replace electrical connector terminals, pins, harnesses, seals, and locks.

The ECM harness electrically connects the ECM to other modules in the vehicle engine and passenger compartments. Wire harnesses should be replaced with proper part number harnesses. When signal wires are spliced into a harness, use wire with high temperature insulation only. With the low current and voltage levels found in the system, it is important that the best possible bond at all wire splices be made by soldering the splices. Molded-on connectors usually require complete replacement of the connector. This means splicing a new connector assembly into the harness. Use care when probing a connector or replacing terminals in them. It is possible to short between opposite terminals. Use jumper wires between connectors for circuit checking. *Never* probe through the weather-pack seals. A connector test adapter kit is used to probe terminals during diagnosis. A fuse remover and test tool should be used when removing a fuse.

When diagnosing, open circuits are often difficult to locate by sight because of oxidation, or because the terminal alignment is hidden by the connectors. Merely wiggling a connector on a sensor in the wiring harness may correct an open circuit condition. This should always be considered when an open circuit or failed sensor is indicated. Intermittent problems may also be caused by oxidized or loose connections. Before making a connector repair be certain of the type of connector. Weather-pack and pull-to-seal connectors look similar but are serviced differently. Some connectors, such as the coolant sensor, use terminals called microconnectors. They are released by inserting a fine pick tool through the front of the connector to disengage a locking tab on the terminal. To remove, bend the tab toward the terminal, pull the terminal and wire out of the connector. To install, insert the tool into a new connector and bend the tab away from the terminal. Molded plugs and connectors are factory fabricated and must be spliced into the harness. Ensure correct splice procedures are used to make a good electrical connection.

"Pull-to-seal terminals. To install a terminal on a wire, the wire is first inserted through the seal and connectors, then the terminal is crimped on the wire and the terminal is pulled back into the connector to seat it in place. To remove this type of terminal:

- Slide the seal back on the wire.

- Use an insertion tool to release the terminal-locking tab.

- Push the wire and terminal out through the connector. If reusing the terminal, reshape the locking tang.

A weather-pack connector can be identified by a rubber seal at the rear of the connector. This connector protects against moisture and dirt that could create oxidation and deposits on the terminals. This protection is important because of the very low voltage and current levels found in the electronic system. To repair a weather-pack terminal, use a special tool to remove the pin and sleeve terminals. If removal is attempted with an ordinary pick, the terminal could be bent or deformed. Unlike standard blade-type terminals, these terminals cannot be straightened once they are bent. Make certain the connectors are properly seated and all of the sealing rings are in place when connecting leads. The hinge-type flap provides a back-up or secondary locking feature for the connector. They are used to improve the connector reliability by retaining the terminals if the small terminal lock tangs are not positioned properly. Weather-pack connections cannot be replaced with a standard connection.

Task F2.8   Connect diagnostic tool to vehicle/engine to access allowed service parameters; determine needed repairs.

Modern vehicle electronic systems electronically store customer options. These are used to change the operational characteristics of the chassis electronic system. To access and reprogram this information, connect a diagnostic tool to the data link, which is usually located under the dash on the driver's side. Examples of customer parameters that can be changed using the appropriate software are: maximum road speed, tire size, axle ratios, driver reward profiles, governor type (limiting speed or variable speed), idle shutdown time, automatic stop/restart, variable horsepower, torque curve, and progressive shifting. These customer parameters or custom calibrations are also able to be copied from one ECM to another if ECM replacement is necessary.

Task F2.9   Use a diagnostic tool (hand-held or PC based and/or break-out cable or box) to inspect and test electronic engine control system, sensors, actuators, electronic control modules (ECMs), and circuits; determine needed repairs.

Accessing information from the vehicle electronic control modules may be accomplished using each manufacturer's specific electronic service tools (ESTs), or by using PC-based software that is available by license agreements from each manufacturer. The Pro-Link 9000, which has become an industry standard, is also widely used. Each manufacturer's ESTs and PC-based software by agreement will access and retrieve information from other manufacturers' products, although some information is regarded as proprietary and so each usually features self-diagnostics that greatly exceed the minimum requirements. When using Pro-Link, make sure that you have the correct cartridge installed. A multiprotocol cartridge is available that accepts a variety of software cartridges for engine, transmission, and brake systems. You will need the correct adapters and multipin connectors to access the system. Most vehicles are equipped with either a 6-pin or 9-pin Deutch SAE/ATA J1587/J1708/J1939 connector, which is usually referred to as the ATA connector. The connector is located in the engine compartment, or in the dash, usually to the left of the steering column. This common connector and the software protocols are established by agreement and enables proprietary software used by various OEMs to communicate with and at least read the parameters and conditions of their competitors systems. Most engine management systems must also communicate with other chassis systems to optimize engine performance. Begin by ensuring that the ignition switch is in the off position. The diagnostic equipment must also be off. Next, connect your diagnostic equipment and turn it on. Turn on the ignition and observe whether communication is established. You will be presented with a series of prompts or menu choices that access different portions of the software. When codes are present, it is usually useful to record the codes, clear them, and retest to see whether they reappear. Codes that reappear indicate an active code that should be repaired before proceeding. Most self-diagnostics will provide a data-stream allowing you to observe the values from each of the sensors and switches that are computer inputs. Any values that are suspected to be in error should be verified using a digital multimeter (DMM). In addition, actuator tests allow you to command the various actuators and observe their operation. For example, in injector solenoid tests that are performed with the key on and the engine off (KOEO), the ECM actuates each injector solenoid in sequence and their operation may be verified because they are heard. This is referred by one manufacturer as a buzz test. Another example is performed with the key on and the engine running (KOER), and is referred to as a cylinder cut-out test. Either one or multiple cylinders, when selected, are cut out and their effect on idle speed and their relative contribution to engine power output is calculated. Similar testing of actuators is available for not only the engine, but also other systems like ABS brake systems. You can make a multitude of tests using a single attachment to the self-diagnostic terminal. Keep in mind that the values that you are observing are those that the computer receives; they are subject to the computers sampling rate, and may be influenced by the integrity of the wiring and connectors that convey them.

Task F2.10   Measure and interpret voltage, voltage drop, amperage, and resistance readings using a digital multimeter (DMM).

Voltage Testing
The voltmeter is considered by many to be the most useful meter available, and they are relatively inexpensive. It is a high resistance meter, so it can be placed across various power sources without damage to the meter. It is further recommended that a high impedance digital voltmeter be used whenever working with vehicles that have electronic modules or computer management systems. High impedance meters will have very high resistance, typically 10 million ohms internal resistance, compared to analog meters having only 10,000 ohms internal resistance. It is important to prevent unintended damage to the system you are testing. Some of the tests that can be performed with a voltmeter are available voltage at the battery, source voltage at virtually any electrical component, to indicate resistance when performing voltage drop testing (explained below). The two test leads are commonly red and black. The red lead is attached to the most positive (higher potential) portion of the circuit and the black lead may be connected to a good ground or to a point on the circuit with lower

voltage (lower potential) than the red lead, depending on the type of test being performed. This places the meter in parallel to the portion of the circuit being tested and is sometimes referred to as a shunt. For the meter to indicate a voltage, the circuit should be turned on and have current flowing. Most modern meters have additional features that make them more useful, like auto ranging, touch hold, and min/max, so learn how to use your meter as effectively as possible. A thorough knowledge of electrical theory will help you understand the meaning of the results of your tests. Study Ohms Law and the principles of series and parallel circuits.

Voltage Drop Testing

Often it is desirable to know what amount of source voltage is being lost or dropped in a circuit by overcoming resistance. This is measured as voltage drop. Voltage drop is the amount of electrical pressure that is lost or consumed pushing current through the circuit. Voltage drop occurs when current passes through various load components and resistances. Most of the voltage should be dropped across the load and not across the components that convey voltage and control the load. Voltage drop testing to measure resistance has several big advantages over using an ohmmeter. First, it is very accurate. You can find very small resistances where voltage is lost (dropped) passing through connections, relays, switches, fuses, wiring, and other voltage conveying components in the circuit. Another advantage is that it is a dynamic test, performed with the circuit live and under load. It is not necessary to isolate the component being tested from the circuit by disconnecting or removing it. Simply connect the voltmeter in parallel with the portion of the circuit being tested, being careful to observe polarity (see figure).

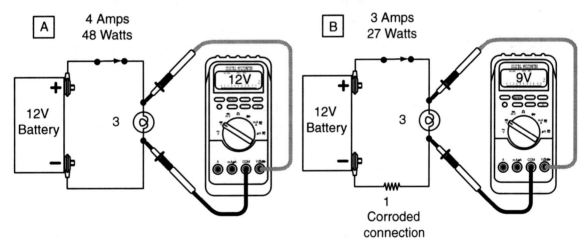

Compare the voltmeter readings above. In the example on the left, almost all source voltage is dropped across the intended load. In the example on the right, only 9 volts are dropped across the intended load and 3 volts are wasted in the rest of the circuit. The load device on the left is operating much more efficiently than the one on the right. By moving the voltmeter leads to various points along the circuit, it is possible to identify where excess voltage is dropped.

Amperage Testing

Amperage testing is different from voltage testing because the meter must be placed in the circuit in series, unless you are using and inductive pickup, which measures current flow by measuring the strength of the magnetic field produced. To use an ammeter, disconnect a wire and connect the meter into the circuit. With the meter in series with the rest of the circuit, all of the current flowing must pass through the meter. If the current exceeds the capacity of the meter, it will be damaged. For this reason, digital meters are usually protected by internal fuses. If current exceeds the fuse rating, it will blow and the meter stops working. Measuring about 10 amps is about the limit of most digital meters. An inductive pickup is useful for measuring larger current flow. A big advantage in using one is that it may be clamped over almost any portion of the circuit without the need to disconnect anything to place the meter in series in the circuit.

Resistance Testing

Resistance is measured in ohms or fractions of an ohm. The ohmmeter is self-powered, and passes a very small current supplied by an internal battery through the circuit being tested. Do not attempt to test a live circuit; that is, one that has voltage in it from any source other than the meters battery. Testing a live circuit can damage an ohmmeter. Several scales are available to permit accurate testing of a wide range of resistance values. Choose the correct range for the task. When a range that is too low is used, the meter may falsely indicate an infinite resistance. Start with the highest range and move to a lower scale if a very low reading is obtained. Auto ranging meters will select the correct scale. It is frequently necessary to remove the component being tested or otherwise isolate it from the circuit to get accurate results.

G. Starting and Charging System Diagnosis and Repair (4 Questions)

Task G1   Perform battery state-of-charge test; determine needed service.

Disconnect the battery cables from the terminals. Install a voltmeter and battery load tester to the posts. Remove the surface charge from recently charged batteries by applying a 300-ampere load across the ports for 15 seconds. The easiest way to load the batteries in the vehicle is to disable the engine from starting, and crank the starter for *no longer* than 10 to 15 seconds. Turn off the load and wait 1 to 2 minutes for the battery to recover. A fully charged battery will have an open circuit voltage of 12.6 volts. An open circuit voltage of 12.4 indicates that the battery is less than fully charged. An open circuit voltage of 12 volts means the battery is nearly completely discharged.

Task G2   Perform battery load and capacitance tests; determine needed service.

Capacity Test

A capacity test one means of checking a battery to determine its ability to function as required in the vehicle. A 12-volt battery that will maintain 9 volts or better during a capacity test is considered a good battery. To perform this test, use equipment that will take a heavy electrical load from the battery, such as a carbon pile or other suitable means.

1. Connect voltmeter, ammeter, and variable load resistance. *Note:* Observe polarity of meters and battery when making connections and select correct meter range.

2. Apply a load to the battery equal to 3 times the ampere-hour rating of the battery for 15 seconds ($3 \times 200 = 600$ amperes).

3. With ammeter reading specified load, read voltage should not be less than 9 volts for a 12-volt battery.

4. If voltmeter shows 9 volts or more, the battery has good output capacity and will readily accept a normal charge. If specific gravity is 1.215 or more, no service is required. If specific gravity is below 1.215, check the charging circuit to determine the cause and correct as required. The battery should be slow-charged for city driving; with highway driving and a good charging system, the battery should charge satisfactorily.

5. If voltmeter shows a reading of less than 9 volts, perform "Battery Light Load Test."

Battery Light Load Test
Check electrical condition of battery cells as follows:

1. If electrolyte level in cells is low, fill to proper level with water and boost charge battery.

2. Place load on battery by cranking engine. If engine starts, turn it off immediately. If engine does not start, hold starter switch on for 3 seconds, then release.

3. Turn on headlights (low beam). After 1 minute, with lights still on read voltage of each cell with voltmeter, compare readings with the following:

   • Good Battery. If all cells read 1.95 volts or more and the difference between highest and lowest cell is less than 0.05 volt, battery is good.

- Good Battery. If cells read both above and below 1.95 volts and the difference between the highest and lowest cell is less than 0.05 volt, battery is good but requires charging. Replace battery if any cell reads 1.95 volts or more and there is a difference of 0.05 volts or more between the highest and lowest cell.
- Discharged Battery. If all cells read less than 1.95 volts, battery is too low to test properly.

Note: Failure of meter to register on all cells does not indicate a defective battery.

4. Boost charge battery and repeat "light load test." If battery is found to be good after boosting, it should be fully recharged for good performance.

5. If none of the cells come up to 1.95 volts after the first boost charge, the battery should be given a second boost.

6. Batteries that do not respond after second boost charge should be replaced.

Voltage and Temperature Chart

TEMPERATURE	MINIMUM VOLTAGE
21°C (70°F and above)	9.6
10°C (50°F)	9.4
–1°C (30°F)	9.1
–10°C (15°F)	8.8
–18°C (0°F)	8.5

Task G3   Charge battery using slow or fast charge method as appropriate.

An open circuit voltage test is used to determine a battery's state of charge. Follow these steps to test voltage. If the battery has just been charged or has recently been in for service, the surface charge must be removed before an accurate voltage measurement can be made. To remove the surface charge, crank the engine for 15 seconds. Do not allow the engine to start. If the battery is out of the vehicle, a load test of half the rated cold cranking amps for 15 seconds will remove the surface charge. After loading the battery by either method, let the battery rest for 1 to 2 minutes. Connect a voltmeter across the battery terminals and observe the reading. If the reading is 12.6 volts or more, the battery should be fully charged. If the voltage reading is below 12.6 volts, the battery probably needs to be charged.

Slow Charging

The "slow charge" method supplies the battery with a relatively low current flow for a relatively long period of time. Only this method will bring the battery to a fully charged state. The "slow charge" method consists of charging at approximately a 9-ampere rate for 24 hours or more, if necessary, to bring the battery to full charge. A fully charged condition is reached when 3 consecutive specific gravity readings taken at hourly intervals, show no increase.

Fast Charging

The "fast charge" method supplies current to the battery at a 40 to 50 ampere rate over a 1 1/2-hour time period. If the electrolyte temperature reaches 125°F before the 1 1/2-hour period is completed, the battery must be taken off charge temporarily or the charging rate reduced to avoid damaging the battery. Although a battery cannot be brought to a fully charged condition during "fast charge," it can be substantially recharged or "boosted." To bring the battery to a fully charged condition, the charging cycle must be finished by the "slow charge" method.

Task G4   ## Start vehicle using jumper cables, a booster battery, or an auxiliary power supply.

If the vehicle will not start due to a discharged battery, using energy from another battery can often start it; this procedure is called "jump starting." Make sure that the boost vehicle also has the same voltage starting system and that it is the negative (−) terminal, which is grounded. Do not attempt to jump-start if you are unsure of the other vehicle's voltage or ground. Diesel engine vehicles have more than one battery because of the higher torque required to start a diesel engine. This procedure can be used to start a single battery vehicle from any of the batteries of the diesel vehicle. However, at low temperatures it may not be possible to start a diesel engine from a single battery in another vehicle.

Position the vehicle with the good (charged) battery so that the booster (jumper) cables will reach, but never let the vehicles touch. Also, be sure the booster cables do not have loose or missing insulation. In both vehicles:

- Turn off the ignition (engine control switch) and all lamps and accessories.

- Apply the parking brake firmly. Shift the transmission to neutral.

- Making sure the cable clamps do not touch any other metal parts, clamp one end of the first booster cable to the positive (+) terminal on one battery, and the other end to the positive terminal on the other battery. Never connect (+) to (−).

- Clamp one end of the second cable to the negative (−) terminal of the good (charged) battery. Make the final connection to the frame rail, chassis, or to any solid, stationary, unpainted metallic object on the engine at least 18 inches (450 mm) from the discharged battery or some other well-grounded point if the battery is mounted outside the engine compartment. Make sure the cables are not on or near pulleys, fans, or other parts that will move when the engine is started.

- Start the engine of the vehicle with the charged battery and run the engine at a moderate speed. Then crank the engine of the vehicle that has the discharged battery. Do not crank for more than 30 seconds.

- Remove the booster cables by reversing the installation sequence. While removing each clamp, take care it does not touch any another metal while the other end remains attached.

Auxiliary power sources: When jump starting from another vehicle is not successful, use a portable generator designed for this use. Exercise care to avoid damage from transient voltage spikes when attempting to start a vehicle equipped with an electronically managed engine. Most electronically managed engines are designed to withstand short-term voltage overloads, but the penalty for excessive overloads can be very costly. Other electronic modules on the vehicle like ABS, Traction Control, and Automatic Temperature Control systems may also suffer damage from voltage spikes. Consideration should be given to charging the batteries or moving the vehicle into a repair facility.

Task G5   ## Inspect, clean, repair or replace batteries, battery cables, disconnects, and clamps.

A battery can be cleaned with a baking soda and water solution. Always wear hand and eye protection when servicing batteries and their components. If the built-in hydrometer indicates light yellow or clear, the electrolyte level is low, and the battery should be replaced. The low electrolyte level may be caused by a high voltage regulator setting that causes overcharging. When disconnecting battery cables, always disconnect the negative battery cable first.

Spray the cable clamps with a protective coating to prevent corrosion. A little grease or petroleum jelly will also prevent corrosion. Also available are protective pads that clip under the clamp and around the terminal to prevent corrosion.

When replacing a battery, check the charging system. A defective alternator or regulator could cause a discharged or overcharged condition, resulting in a damaged battery. The cables should be inspected and the terminals cleaned or replaced as part of a good PM routine.

When removing a battery from a vehicle, the battery ground cable is always removed first and the positive cable last. The order is reversed when installing the battery and the cables.

Task G6 **Inspect, test, and reinstall or replace starter relays, safety switch(s), and solenoids.**

Some starter relays are integral with the starter motor and are replaced as a unit. Other starter relays can attach either to the starter motor as a separate unit or to the firewall. Removal of these types of starter relays entails removing the two positive leads and the control wires. The starter relay is an electrical high current switch, and as such, it does not have a ground lead. Any time maintenance is to be performed on the starter, remove the ground (negative lead) from the battery to prevent accidental shorting of the wiring.

Task G7 **Perform alternator voltage and amperage output tests; determine needed repairs.**

Testing the alternator output voltage and amperage is easily performed. If a battery load tester is available, connect it across the battery, observing correct polarity. Start the engine and increase to at least 1500 rpms. Observe the unloaded output voltage; it should be between 13.5 to 14.5 volts. Attach an amp clamp around the alternator output wire. With the engine at high idle, use the load tester to reduce battery voltage to obtain the highest amperage reading without causing the voltage to fall below 12 volts. This procedure forces the alternator to increase its output to near its maximum. The amp clamp should record output within 5 percent of the alternators capacity. If a load tester is not available, turning on as many electrical loads as possible may be used as a substitute. If an undercharge is noted, some of the causes are: loose or glazed belts, damaged or corroded wires and connectors, a defective battery, a defective AC generator, a defective diode trio, or a faulty voltage regulator. Follow the relevant diagnostic procedures to find the specific fault.

Task G8 **Perform starter and charging circuit voltage drop tests; determine needed repairs.**

It is valuable to know how much battery voltage is dropped in the starting circuit overcoming resistance. This may be accomplished by voltage drop testing. Voltage drop is the amount of electrical pressure that is lost or consumed pushing current through the circuit. Voltage drop occurs when current passes through load components and resistances. Most of the voltage should be dropped across the starter and not across the cables and wiring that conduct voltage and control the starter. Voltage drop testing to measure resistance has several big advantages over using an ohmmeter. First, it is very accurate. You can find very small resistances where voltage is lost (dropped), passing through connections, relays, switches, fuses, wiring, and other voltage conveying components in the circuit. Second, it is a dynamic test, performed with the circuit live and under load. It is not necessary to isolate the component being tested from the circuit by disconnecting or removing it. Simply connect the voltmeter in parallel with the portion of the circuit being tested, being careful to observe polarity. For example, to test for resistance caused by battery terminal corrosion, place one voltmeter lead on the post and the other one on the cable, and then observe the meter while cranking the starter. A good connection will show that only about 0.1 V is dropped; this is for only the one connection. When the voltage drop across all of the positive battery cable is measured, the reading should be less than 0.5 volts. You can test the positive battery cable by placing one lead on the battery post and the other lead on the solenoids battery cable stud, thus testing the entire cable. A similar test with one lead on the negative battery post and the other on the starter frame will test the entire ground side of the circuit. There should be less than 0.3 V dropped across the ground circuit. Most of the battery voltage should be dropped across the starter motor. About 0.3 V is dropped across the starter solenoid and the remainder is consumed elsewhere in the circuit.

Voltage drop testing is also useful to find unwanted resistance in the charging system. The tests are similar to those performed on the starting system. Most of the voltage is dropped when the alternator is producing its maximum output, so turn on as many electrical loads as possible. From the B+ terminal of the alternator to the positive battery post, voltage drop should be less than 0.2 volts. Voltage drop from the ground terminal on the alternator to the negative battery post should be less than 0.3 volts.

5

Sample Test for Practice

Sample Test

Please note the letter and number in parentheses following each question. They match the task in Section 4 that discusses the relevant subject matter. You may want to refer to the overview using the cross-referencing key to help with questions posing problems for you.

1. A vehicle being fueled continuously has the fuel nozzle click off prematurely. Which of these is the Most-Likely cause?
   A. a blocked tank vent valve
   B. water in the fuel tank
   C. plugged fuel filters
   D. fuel temperature is too low   (F1.1)

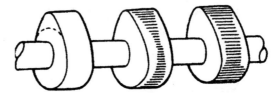

2. Two checks of the heel-to-toe measurement on the camshaft in the figure provide different readings, but they are still within specification. Technician A says to check for grooving of the cam lobe. Technician B says that providing the readings are within specifications no further checks are required. Who is correct?
   A. A only
   B. B only
   C. Both A and B
   D. Neither A nor B   (C7)

3. Which of the following would Most-Likely indicate oil cooler failure?
   A. low oil pressure
   B. high oil pressure
   C. milky gray sludge in radiator or surge tank
   D. high engine oil level   (D4)

4. Technician A says that valve crossheads should be magnafluxed to check for cracks. Technician B says the valve crosshead guide pins should be checked for squareness. Who is correct?
   A. A only
   B. B only
   C. Both A and B
   D. Neither A nor B   (B6)

5. All of the following are steps in an air inlet restriction test EXCEPT:
   A. Connect a manometer to the air intake downstream from the filter.
   B. Remove the air cleaner element and duct.
   C. Check the manometer reading with the engine under load.
   D. Record the inlet restriction spec in inches of water.   (A7)

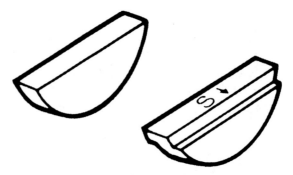

6. The figure shows keys used on camshaft gears. Technician A says the key on the right is used to offset the timing and it must be replaced with the same type. Technician B says that the camshaft, gear, and key must be replaced because the camshaft Most-Likely jammed and started to shear the key. Who is correct?
   A. A only
   B. B only
   C. Both A and B
   D. Neither A nor B   (B11)

7. Upon inspection of the engine compartment, a technician finds wiring connectors hanging loose in the engine compartment. The technician should
   A. wire tie the loose connectors out of the way.
   B. cut the connectors off and heat-shrink the wires.
   C. place dummy plugs over the open connector ends.
   D. consult appropriate service manuals to identify where the connectors belong.   (A17)

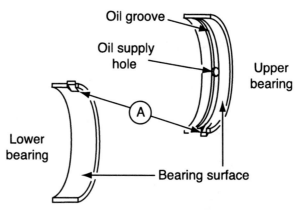

8. The notch in the bearing shell at point A in the figure is used to:
   A. direct the oil evenly around the bearing.
   B. indicate a defect in the bearing shell.
   C. indicate this is an oversized bearing shell.
   D. position the bearing shell in the bearing bore.   (C9)

9. Technician A says purging air from the fuel system can be done with the hand primer. Technician B says purging (bleeding) air from the fuel system should be done any time a fuel hose is replaced. Who is correct?
   A. A only
   B. B only
   C. Both A and B
   D. Neither A nor B   (F1.3)

10. An engine cranks slowly and fails to start. Technician A says that the ambient temperature may be too low to achieve cylinder combustion temperature. Technician B says to check for low battery potential or corrosion on the battery terminals and clean and charge as necessary. Who is correct?
    A. A only
    B. B only
    C. Both A and B
    D. Neither A nor B   (A11)

11. When removing electronic unit injectors (EUIs) from a diesel engine cylinder head, Technician A says that the fuel rail should be drained before pulling the first EUI to avoid rail fuel from pouring into the cylinder. Technician B says that after changing a set of EUIs, new calibration codes must be programmed to the engine ECM or a cylinder balance problem could result. Who is correct?
    A. A only
    B. B only
    C. Both A and B
    D. Neither A nor B   (F2.4)

12. Technician A says that when an oil pump is disassembled, all mated surfaces must be checked for wear. Technician B says because oil pump gears receive so much wear, the oil pump should routinely be replaced during an overhaul. Who is correct?
    A. A only
    B. B only
    C. Both A and B
    D. Neither A nor B   (D2)

13. A vibration that has an occurrence rate equal to engine speed is LEAST-Likely to be caused by a(n):
    A. out-of-balance turbocharger.
    B. bent connecting rod.
    C. out-of-balance crankshaft.
    D. dented vibration damper.   (A13)

14. A vehicle's battery is being tested. Technician A says slow charging is the best method to fully charge a battery. Technician B says the surface charge should be removed before testing the battery. Who is correct?
    A. A only
    B. B only
    C. Both A and B
    D. Neither A nor B   (G3)

15. In electronic diagnostics the acronym PID stands for
    A. parameter identifier.
    B. part identification number.
    C. positive induction device.
    D. partial induction delivery.   (F2.1)

16. When checking the coolant condition with an SCA test strip, the technician finds that the coolant is *extremely* overconditioned. What should the technician do?
    A. Add more antifreeze to increase the SCA.
    B. Continue to run the bus until the next PMI.
    C. Drain the entire coolant system and add the proper SCA mixture.
    D. Run the bus with no SCA additives until the next PMI.   (D11)

17. A diesel engine has low power and overheats. Technician A says the problem may be a particulate filter that needs to be serviced. Technician B says the problem may be a stuck closed exhaust brake. Who is correct?
    A. A only
    B. B only
    C. Both A and B
    D. Neither A nor B   (E5)

18. The firing order of an in-line, 6-cylinder, 4-stroke cycle diesel engine is 153624. Which of the following would be correctly described as companion cylinders?
    A. 1 and 5
    B. 4 and 5
    C. 3 and 6
    D. 5 and 2   (B10)

19. A diesel engine has excessive crankcase pressure and is leaking oil from several crankcase seals. Technician A says crankcase pressure can increase if the crankcase breather tube is restricted. Technician B says that excessive blowby increases peak engine power. Who is correct?
    A. A only
    B. B only
    C. Both A and B
    D. Neither A nor B   (A10)

20. Technician A says that glow plugs can be used to help start the engine in cold weather. Technician B says that most modern large diesel engines use glow plugs to start engines in cold weather. Who is correct?
    A. A only
    B. B only
    C. Both A and B
    D. Neither A nor B   (E6)

21. Which of the following vehicle programming categories would be LEAST-Likely to be classified as customer data programming?
    A. maximum road speed
    B. fuel map
    C. driver rewards
    D. governor type (LS or VS)   (F2.8)

22. Excess exhaust back pressure may result in all of the following EXCEPT:
    A. lower engine power
    B. higher exhaust temperature
    C. higher intake restriction readings
    D. poor combustion   (A9)

23. A diesel engine will start and run fine first thing in the morning. However, after the engine runs for about 2 hours, it will die. When the operator attempts to restart, the engine will start and die repeatedly. After the engine is allowed to sit for a couple of hours, it will start and run fine again. Technician A says the problem may be a faulty pull in the windings in the shutdown solenoid. Technician B says the problem may be a faulty hold in the windings on the shutdown solenoid. Who is correct?
   A. A only
   B. B only
   C. Both A and B
   D. Neither A nor B   (F1.11)

24. Which of the following is the correct tool to use to check flywheel housing alignment?
   A. alignment machine
   B. micrometer
   C. dial indicator
   D. straightedge and feeler gauge   (C17)

25. When replacing a starter relay, which of the following would LEAST-Likely be removed or disconnected?
   A. positive lead to the relay
   B. negative lead to the relay
   C. relay mounting nuts
   D. ground cable from battery   (G6)

26. When using jumper cables to start a vehicle with a dead battery, always do which of these items?
   A. Connect (+) positive to (–) negative to create a series aiding circuit.
   B. Connect (+) positive to (+) positive and (–) negative to (–) negative to create a parallel circuit.
   C. Connect the (–) negative cable first and the (+) positive last.
   D. Electronically ground both vehicles before making any connections.   (G4)

27. When inspecting an oil filter housing or its mounting, which of the following is LEAST-Likely to be a procedure?
   A. visually check for cracks
   B. visually check gasket surface for nicks
   C. magnaflux the housing to detect small cracks
   D. visually inspect housing passageways for obstructions   (D3)

28. A misfiring cylinder is diagnosed on a diesel engine with mechanical unit injectors. The technician replaces the unit injector on the misfiring cylinder. However, the engine continues to miss. Which of these could be the cause?
   A. a worn cam follower roller
   B. a restricted fuel filter
   C. high fuel pump pressure
   D. buffer screw adjustment   (A12)

29. When diagnosing EGR-related engine problems, Technician A says vehicle air system pressure should be checked. Technician B says the ECM should be checked for diagnostic trouble codes (DTCs). Who is correct?
   A. A only
   B. B only
   C. Both A and B
   D. Neither A nor B   (E7)

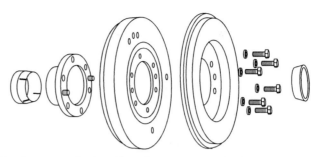

30. Which of the following properly describes the component in the figure?
    A. flywheel assembly
    B. vibration damper assembly
    C. timing gear assembly
    D. water pump assembly   (C16)

31. Technician A says a nonroller (flat) lifter base should be smooth and slightly concave.
    Technician B says if a lifter base is pitted, the camshaft mating lobe should be inspected. Who
    is correct?
    A. A only
    B. B only
    C. Both A and B
    D. Neither A nor B   (B9)

32. To inspect a cylinder head for cracks, all of the following may be necessary EXCEPT:
    A. Sand or peen the cylinder head to remove excess gasket material from the head and inspect
    for visible cracks between cylinders and ports.
    B. Check the torque of head bolts.
    C. Use magnetic crack detection to find small cracks.
    D. Remove excess carbon build-up.   (B3)

33. The removal of an injector from a cylinder head is being discussed. Technician A says you can
    use a slide hammer to remove the injector. Technician B says you can also use a puller to
    remove the injector. Who is correct?
    A. A only
    B. B only
    C. Both A and B
    D. Neither A nor B   (F1.8 and F2.3)

34. A mechanical unit injector fueled engine appears to be misfiring. Technician A says to use an
    infrared thermometer to find the dead cylinder. Technician B says to manually cancel out each
    injector and locate the one that does not affect engine performance. Who is correct?
    A. A only
    B. B only
    C. Both A and B
    D. Neither A nor B   (A4)

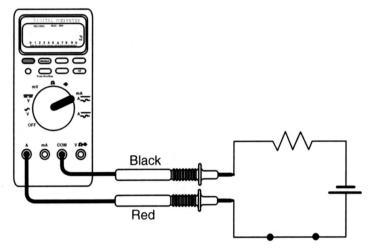

35. In the figure, what is about to be measured in the circuit shown?
    A. current flow
    B. voltage drop
    C. circuit resistance
    D. wattage consumed   (G7)

36. Technician A says that cupping of the cylinder valve head is normal and should not be a reason
    to replace a valve. Technician B says that valve margin measurements need not be taken for
    intake valves. Who is correct?
    A. A only
    B. B only
    C. Both A and B
    D. Neither A nor B   (B4)

37. Technician A says bent valvetrain pushrods and tubes should be replaced, not straightened.
    Technician B says pushrods or tubes must be checked for ball socket integrity. Who is correct?
    A. A only
    B. B only
    C. Both A and B
    D. Neither A nor B   (B8)

38. A diesel engine produces heavy black smoke upon acceleration. Otherwise engine operation is
    normal. Technician A says the problem may be a restricted fuel filter. Technician B says the
    problem may be a restricted air filter. Who is correct?
    A. A only
    B. B only
    C. Both A and B
    D. Neither A nor B   (F1.9)

39. Technician A says that a radiator pressure cap with a 9 stamped on it should open to release
    pressure at between 8 psi (55 kPa) to 10 psi (69 kPa). Technician B says the same cap should
    open at about 0.6 psi (4.3 kPa) differential (vacuum) pressure. Who is correct?
    A. A only
    B. B only
    C. Both A and B
    D. Neither A nor B   (D13)

40. Technician A says end gaps on piston rings must be offset to limit blowby. Technician B says when installing a piston and connecting rod, studs should be covered to prevent damage to the crankshaft. Who is correct?
    A. A only
    B. B only
    C. Both A and B
    D. Neither A nor B   (C15)

41. A water pump mounting hole in the cylinder block has stripped threads. Technician A says that the block has to be replaced. Technician B says it may be possible to install a helicoil. Who is correct?
    A. A only
    B. B only
    C. Both A and B
    D. Neither A nor B   (B2)

42. A light film of slippery liquid with an antifreeze odor covers the top of most engine compartment components. Technician A says that the radiator may have a pinhole leak. Technician B says that the upper radiator hose may be leaking. Who is correct?
    A. A only
    B. B only
    C. Both A and B
    D. Neither A nor B   (A2)

43. A diesel engine has bubbles flowing in the clear bowl on the fuel\water separator. Technician A says this is a normal condition. Technician B says this is caused by leaks on the suction side of the fuel system. Who is correct?
    A. A only
    B. B only
    C. Both A and B
    D. Neither A nor B   (A6)

44. The LEAST-Likely cause for a low oil pressure condition is
    A. a stuck closed oil pressure relief valve.
    B. a restricted oil pump suction tube.
    C. worn main bearings.
    D. a clogged oil filter.   (A15)

45. A manifold pressure test is being performed. Technician A says you may use an Hg manometer or pressure gauge to check pressure. Technician B says the engine must not be operated at speeds above idle. Who is correct?
    A. A only
    B. B only
    C. Both A and B
    D. Neither A nor B   (A8)

46. All of these statements about mechanical unit injectors (MUI) are true EXCEPT:
    A. They receive fuel at charging pressure from a gear pump.
    B. Engine output is controlled by a mechanical governor.
    C. Injector pulse width is controlled by the injector driver module (IDM).
    D. Fuel metering is controlled by the position of the rack.   (F1.7)

47. Technician A says to clean and blow out the seal groove in a rocker housing cover with compressed air. Technician B says when using silicone for the valve cover gasket a light coat of the oil in the seal groove makes the gasket easier to install. Who is correct?
    A. A only
    B. B only
    C. Both A and B
    D. Neither A nor B   (C1)

48. Cracks are present in the piston skirt. Technician A says to check the connecting rod and bottom of the cylinder liner for damage caused by improperly installed connecting rods. Technician B says to check for insufficient piston-to-cylinder liner clearance. Who is correct?
    A. A only
    B. B only
    C. Both A and B
    D. Neither A nor B   (C11)

49. The results of testing with SCA test strips can indicate all of the following EXCEPT:
    A. coolant exposure to combustion gases
    B. improper use of a low silicate antifreeze
    C. coolant temperature
    D. coolant degradation   (D8)

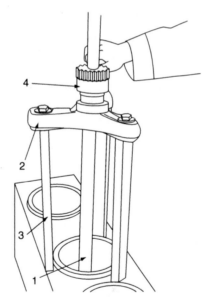

50. As shown in the figure, Technician A says puller adapter plates or shoes must fit snugly in sleeves to prevent cocking in the block. Technician B says if puller adapter plates are larger than the sleeve OD they are better because they cannot slip. Who is correct?
    A. A only
    B. B only
    C. Both A and B
    D. Neither A nor B   (C5)

51. Technician A says that petroleum jelly can be used on battery cable clamps to prevent corrosion. Technician B says it is permissible to use protective pads and a little grease to prevent corrosion on battery cable clamps. Who is correct?
    A. A only
    B. B only
    C. Both A and B
    D. Neither A nor B   (G5)

52. A transit vehicle is being checked for a low power complaint, along with high engine temperatures. Which of these could be the Most-Likely cause?
    A. a clogged air filter
    B. a stuck closed exhaust brake valve
    C. aerated fuel
    D. a hydraulic fan stuck on high speed   (E8)

53. Technician A says if the pressure in the radiator is too great, the pressure cap will relieve the excess pressure and fluid to an expansion tank. Technician B says most radiator caps have a release pressure of 15 psi or lower. Who is correct?
    A. A only
    B. B only
    C. Both A and B
    D. Neither A nor B   (D15)

54. An engine with an electronic fuel system has recorded an input circuit fault. Technician A says to use only a digital multimeter to verify voltages and sensor values. Technician B says that a multimeter used to diagnose engine electronics should have at least 10-mega ohm impedance. Who is correct?
    A. A only
    B. B only
    C. Both A and B
    D. Neither A nor B   (F2.6)

55. An operator is concerned with poor fuel economy. The technician notices that the exhaust smell is strong at idle and there is a slight amount of smoke on acceleration. There are no trouble codes. Technician A says the turbo boost sensor could be reading a slight amount of turbo boost when there is none. Technician B says the throttle position sensor could be reading slight throttle input when there is none. Who is correct?
    A. A only
    B. B only
    C. Both A and B
    D. Neither A nor B   (E3)

56. Technician A says that all gears in the timing gear train must be inspected for tooth wear, including the crankshaft gear. Technician B says a slight roll or lip on each gear tooth is acceptable, because gears will normally mate this way during the first 500 miles of operation. Who is correct?
    A. A only
    B. B only
    C. Both A and B
    D. Neither A nor B   (C10)

57. A bus is being checked for low fuel pressure. Where is the best place to attach a gauge to obtain an accurate reading?
    A. at the suction filter housing
    B. at the fuel pump outlet
    C. at the fuel tank
    D. at the fuel pump inlet   (F1.4)

58. Which of the following is LEAST-Likely to present a problem in an electrical/electronic circuit?
    A. loose or improperly mated connector
    B. wires with nicks or lumps in the insulation
    C. a melted or distorted electrical connector or wire
    D. ground wires attached to an unpainted frame surface   (A3)

59. Surges in coolant level could be caused by all of the following EXCEPT:
    A. a loose belt
    B. a defective fuel injector
    C. a restricted radiator
    D. a blown head gasket   (A14)

60. To determine battery potential, which measuring instrument is used?
    A. refractometer
    B. voltmeter
    C. ohmmeter
    D. ammeter   (G1)

61. Technician A says a flywheel ring gear is heat-shrink fitted to the flywheel. Technician B says that if ring gear teeth are damaged, the flywheel must be replaced. Who is correct?
    A. A only
    B. B only
    C. Both A and B
    D. Neither A nor B   (C18)

62. When checking for excess resistance in the starting circuit, Technician A says to use an ohmmeter to check the resistance between (–) negative and ground while cranking the starter. Technician B says to check the voltage drop between battery positive and the starter terminal stud while the starter is cranking. Who is correct?
    A. A only
    B. B only
    C. Both A and B
    D. Neither A nor B   (G8)

63. Overfilling of the engine oil can cause all of the following EXCEPT:
    A. engine overheating
    B. aerated oil
    C. high oil pressure
    D. low power complaint   (D6)

64. Which of the following is LEAST-Likely to cause a cooling system failure?
    A. a water pump drive belt having two cracks within any 2-in. section
    B. a drop of coolant from the pump weep hole
    C. a loose clamp on a heater hose
    D. rust or scum floating in the radiator   (D12)

65. To check a cylinder block deck for warpage, Technician A says to use a straightedge and feeler gauge. Technician B says to measure the deck at the 3, 6, 9, and 12 o'clock positions around each cylinder using a dial indicator. Who is correct?
    A. A only
    B. B only
    C. Both A and B
    D. Neither A nor B   (C2)

66. An Electronic Unit Injector diesel engine dies while running down the road and will not restart. No smoke is evident while cranking, and the engine cranks at the normal rpm. Which of the following is the Most-Likely cause?
    A. failed engine speed sensor
    B. stuck open wastegate
    C. clogged air filter
    D. failed engine temperature sensor   (F2.5)

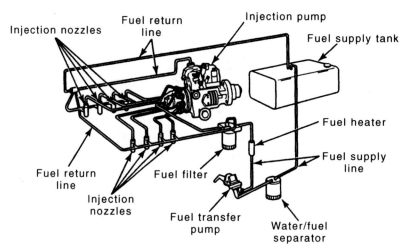

67. What type of fuel system is shown in the figure?
    A. distributor pump
    B. HEUI
    C. in-line pump
    D. unit injector   (F1.6)

68. A diesel engine fails to reach operating temperature. Technician A says a leaking thermostat seal could be the cause. Technician B says a loose fan drive belt could be the cause. Who is correct?
    A. A only
    B. B only
    C. Both A and B
    D. Neither A nor B   (D9)

69. When changing weatherproof electronic connectors, the technician should be careful to
    A. use the insertion/removal tool.
    B. bend the pins to fit.
    C. replace with standard connectors.
    D. ground all wiring before changing.   (F2.7)

70. On startup, the exhaust produces a white/light gray haze for about 15 to 20 seconds, then turns almost clear. Technician A says that the engine may have a broken injector spring. Technician B says this is normal, but advises the operator that it is good practice to use an immersion heater when ambient temperature falls below freezing. Who is correct?
    A. A only
    B. B only
    C. Both A and B
    D. Neither A nor B   (A5)

71. While checking the oil level in the engine the technician notices evidence of coolant in the engine oil. All of the following could cause this EXCEPT:
    A. a crack in the exhaust valve seat
    B. a crack in the water jacket in the engine block
    C. an internal leak in the water pump
    D. a leak in the head gasket   (B1)

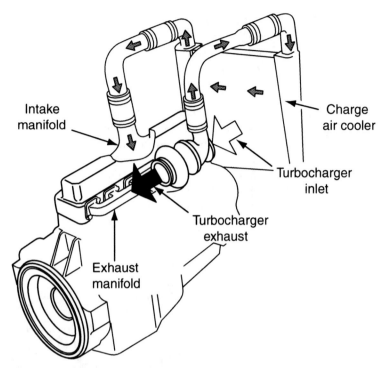

Intake manifold

Charge air cooler

Turbocharger inlet

Turbocharger exhaust

Exhaust manifold

72. Referencing the figure, what cools the boost air?
    A. engine coolant in the intercooler
    B. ram air through the charge-air cooler
    C. filtered intake air from the turbocharger
    D. exhaust gas routed through the charge-air cooler   (E4)

73. Technician A says that excessive smoke from a diesel engine is a good indication that the engine needs to be overhauled. Technician B says discussing the engine operation with the operator can help in diagnosing the cause of a smoking condition. Who is correct?
    A. A only
    B. B only
    C. Both A and B
    D. Neither A nor B   (A1)

74. Which of the following tools is LEAST-Likely to reveal an out-of-round cylinder sleeve?
    A. inside micrometer
    B. telescoping gauge
    C. feeler gauge
    D. dial bore gauge   (C3, C4, and C5)

75. Technician A says back-flushing the radiator in the vehicle should be accomplished annually. Technician B says air locks usually occur when the cooling system is filled with the bleeder valves open. Who is correct?
    A. A only
    B. B only
    C. Both A and B
    D. Neither A nor B   (D10)

76. Cracks and scoring in the piston skirts are LEAST-Likely to be caused by which of the following?
    A. high crankcase lubrication levels
    B. improper piston clearance
    C. overfueling
    D. engine overheating   (C12)

77. Which of the following would be the best source for the most current service information?
    A. technical service bulletins
    B. service manuals
    C. parts manuals
    D. operators manuals   (A18)

78. Technician A says when honing a cylinder liner, the hone should be rhythmically stroked up and down to provide a good crosshatch pattern. Technician B says once the honing has been completed, the liner should be washed with soap solution to remove stone grit. Who is correct?
    A. A only
    B. B only
    C. Both A and B
    D. Neither A nor B   (C4)

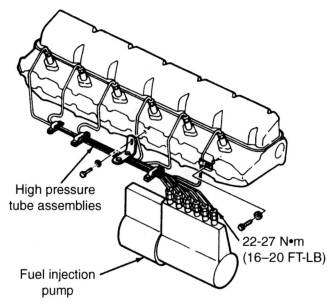

High pressure
tube assemblies

22-27 N•m
(16–20 FT-LB)

Fuel injection
pump

79. In the injector fuel lines in the above figure, Technician A says they must be removed and replaced from the engine as an assembly. Technician B says that it is possible to remove only a line that needs to be replaced. Who is correct?
    A. A only
    B. B only
    C. Both A and B
    D. Neither A nor B   (F1.10)

80. An engine with a mechanical fuel system is reported to have low power. Which of the following components should be checked *first*?
    A. fuel tank crossover pipe restriction
    B. throttle arm break over setting
    C. accelerator linkage
    D. governor settings   (F1.5)

81. A driver is concerned that the engine brake will not disengage when the throttle is depressed. Which of the following is the Most-Likely cause?
    A. clutch micro switch
    B. engine brake master switch
    C. service brake switch
    D. throttle position sensor   (F2.2)

82. The vanes of a turbocharger compressor wheel are excessively worn. Technician A says this could be a result of a loose or damaged hose between the air filter and the turbo. Technician B says that the aftercooler is not properly cooling the air before it enters the turbocharger, causing the compressor blades to become fatigued. Who is correct?
    A. A only
    B. B only
    C. Both A and B
    D. Neither A nor B   (E2)

83. A diesel engine develops too much manifold boost at high engine loads. Technician A says the air filter is punctured. Technician B says the wastegate could be malfunctioning. Who is correct?
    A. A only
    B. B only
    C. Both A and B
    D. Neither A nor B   (E1)

84. All of the following will cause uneven wear on the main bearings or stress fractures on the crankshaft EXCEPT:
    A. full load operation in excess of 8 hours
    B. unbalanced vibration damper
    C. flywheel housing incorrectly installed
    D. loose torque converter bolts   (C14)

85. Technician A says to check the bearings and crankshaft before cleaning the crank. Technician B says to measure the crankshaft after cleaning. Who is correct?
    A. A only
    B. B only
    C. Both A and B
    D. Neither A nor B   (C8)

86. Technician A says that on a turbocharged diesel engine, the turbocharger is lubricated by engine oil. Technician B says that if the turbo shaft bearing fails, the engine oil must be changed. Who is correct?
    A. A only
    B. B only
    C. Both A and B
    D. Neither A nor B   (D5)

87. A single-acting plunger, fuel transfer pump can be equipped with all of the following EXCEPT:
    A. sediment bowl
    B. sediment strainer
    C. hand primer
    D. pop-off valve   (F1.2)

88. Technician A says that the road speed sensor on a typical vehicle electronic management system uses a permanent magnet signal generator to signal road speed data. Technician B says that the road speed sensor is usually located in the tail shaft of the transmission. Who is correct?
    A. A only
    B. B only
    C. Both A and B
    D. Neither A nor B   (F2.9)

89. Technician A says that when two or more identical belts are used on the same pulley, damage on one belt requires that both belts be changed. Technician B says belts can be installed by starting them on the edge of the pulley and cranking the engine. Who is correct?
    A. A only
    B. B only
    C. Both A and B
    D. Neither A nor B   (D7)

90. Technician A says worn camshaft bushings can cause low oil pressure. Technician B says camshaft bushings can be reinstalled after being removed with the correct driver. Who is correct?
    A. A only
    B. B only
    C. Both A and B
    D. Neither A nor B   (C6)

91. When using a voltmeter to perform a voltage drop test in a circuit, how should the test leads be connected?
    A. to the battery terminals
    B. from the positive battery terminal to chassis ground
    C. in series with the circuit being tested
    D. in parallel with the circuit being tested   (F2.10)

92. Technician A says the piston ring end gap should be checked on the piston. Technician B says the piston ring end gap is checked with only the rings in the cylinder. Who is correct?
    A. A only
    B. B only
    C. Both A and B
    D. Neither A nor B   (C13)

93. When performing a cylinder head leakage test of fuel passages, which of the following is LEAST-Likely to be used?
    A. test injectors
    B. compressed air
    C. threaded plug
    D. injection pump   (B5)

94. Which of the following components will LEAST-Likely be found on a HEUI fuel system?
    A. transfer pump
    B. unit injector
    C. hydro mechanical governor
    D. plunger pump   (F2.3)

95. An engine drops a valve 20 hours after an overhaul. Technician A says the probable cause was missing or defective valve seals. Technician B says the probable cause was improperly installed keepers. Who is correct?
    A. A only
    B. B only
    C. Both A and B
    D. Neither A nor B   (B7)

96. Technician A says to replace the fan if the blades are cracked. Technician B says bent fan blades can cause engine vibration. Who is correct?
    A. A only
    B. B only
    C. Both A and B
    D. Neither A nor B   (D14)

97. Which of the following SAE acronyms is used to identify a major chassis electronic system with processing capability such as the engine or transmission?
    A. MID—message identifier
    B. SID—subsystem identifier
    C. PID—parameter identifier
    D. FMI—fault mode indicator   (A16)

98. As shown in the chart, what would be the minimum allowable voltage during a battery load test if the temperature is 50°F (10°C)?
    A. 9.6
    B. 9.4
    C. 9.1
    D. 9.8   (G2)

99. An electric oil pressure gauge indicates 0 pressure, but the engine is well lubricated. Which of the following is Most-Likely at fault?
    A. The oil passage to the sensor is open.
    B. The wrong engine oil was installed.
    C. The wiring to the sensor is open.
    D. The engine is not at operating temperature.   (D1)

6 Additional Test Questions for Practice

Additional Test Questions

Please note the letter and number in parentheses following each question. They match the task in Section 4 that discusses the relevant subject matter. You may want to refer to the overview using the cross-referencing key to help with questions posing problems for you.

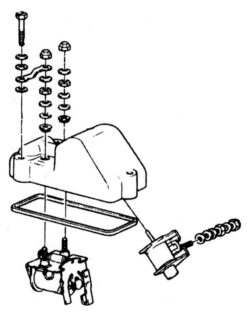

1. In the internal shutdown solenoid in the figure, Technician A says that it must be tested after replacement. Technician B says to clean the old gasket off the cover and fuel pump when changing the shutdown solenoid. Who is correct?
   A. A only
   B. B only
   C. Both A and B
   D. Neither A nor B   (F1.6)

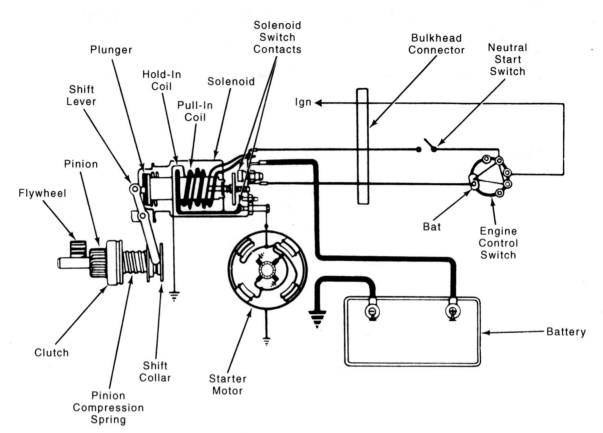

2. When performing voltage drop tests on the starting circuit in the figure, what should be the normal voltage drop across the solenoid switch contacts?
   A. 3.0 volts
   B. More than 9.6 volts
   C. Less than 0.01 volts
   D. 0.3 volts   (G8)

3. When performing failure analysis on a set of crankshaft main bearings, one is observed to be heat discolored on the backing (block) side. Which of the following would be Most-Likely to create this condition?
   A. excessive big endplay
   B. insufficient bearing crush
   C. belled crankshaft main journal
   D. degraded engine lube   (C9)

4. Which of the following is LEAST-Likely to be a symptom of a mechanical governor problem?
   A. random engine miss
   B. high idle overrun
   C. low idle underrun
   D. a hunting condition   (F1.6)

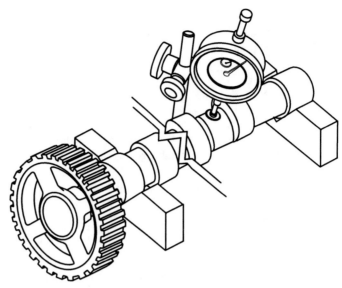

5. What is being indicated on the camshaft in the figure?
   A. intake cam lift
   B. exhaust cam lift
   C. camshaft endplay
   D. cam main journal runout   (B11)

6. While discussing a turbocharged diesel engine, Technician A says that an exhaust manifold leak will affect the performance of the engine. Technician B says that leaks in the intake pipe can cause damage to the turbo impellor. Who is correct?
   A. A only
   B. B only
   C. Both A and B
   D. Neither A nor B   (A12)

7. Technician A says that when replacing a wet-type sleeve, you should always replace the O-rings. Technician B says that new O-rings should be coated with anaerobic silicone. Who is correct?
   A. A only
   B. B only
   C. Both A and B
   D. Neither A nor B   (C5)

8. Technician A says that excessive crankcase pressure can be caused by a plugged oil breather. Technician B says that all diesel engines use exhaust gas recirculation (EGR). Who is correct?
   A. A only
   B. B only
   C. Both A and B
   D. Neither A nor B   (A10)

9. Technician A says to use a digital voltmeter to test grounds with an energized circuit. Technician B says to use an ammeter to test grounds when power is not applied. Who is correct?
   A. A only
   B. B only
   C. Both A and B
   D. Neither A nor B   (F2.10)

10. Technician A says that EUI calibration codes are indicated on each EUI and may be 2, 3, or 4 numeric digits depending on the manufacturer. Technician B says that calibration codes must be programmed to the ECM each time EUIs are replaced. Who is correct?
   A. A only
   B. B only
   C. Both A and B
   D. Neither A nor B   (F2.4)

11. Technician A says that manufacturers allow for relative movement between the chassis and the engine by using durable rubber couplings in the intake system. Technician B says that if an air intake pipe, tube, or hose is loose or missing a clamp, the turbocharger compressor should be inspected. Who is correct?
   A. A only
   B. B only
   C. Both A and B
   D. Neither A nor B   (E1)

12. All of the following are correct procedures if a pressure cap is found to be defective EXCEPT:
   A. replace the cap
   B. test the coolant
   C. check the radiator hoses
   D. adjust or replace the spring in the cap   (D13)

13. Technician A says a loss of engine lubrication oil through the breather tube is an indication of excess crankcase pressure. Technician B says any measurable crankcase pressure is a sign of trouble. Who is correct?
   A. A only
   B. B only
   C. Both A and B
   D. Neither A nor B   (A10)

14. Technician A says pistons should be laid out in order to facilitate failure analysis. Technician B says installing an improper injector nozzle could cause piston failure. Who is correct?
   A. A only
   B. B only
   C. Both A and B
   D. Neither A nor B   (C11)

15. When performing a valve adjustment on a diesel engine with valve bridges, Technician A says the valve bridges must be adjusted before the valve clearance is set. Technician B says the valve bridge adjustment locknuts should torqued with the bridges secured in a vice. Who is correct?
   A. A only
   B. B only
   C. Both A and B
   D. Neither A or B   (B6)

16. During a check of the cooling system on a transit bus, a technician finds the system will not maintain pressure. No coolant leaks are visible anywhere on the vehicle. Technician A says the pressure relief valve may be stuck closed. Technician B says the engine must be running when pressurizing the cooling system. Who is correct?
   A. A only
   B. B only
   C. Both A and B
   D. Neither A nor B   (D15)

17. A small crack is detected in a wet sleeve around the counterbore flange. Technician A says to use helia-arc welding to seal the crack. Technician B says the block needs to have the counterbore recut. Who is correct?
    A. A only
    B. B only
    C. Both A and B
    D. Neither A nor B   (C4)

18. An electronically managed diesel engine produces white smoke on cold startup for at least 10 minutes. Technician A says this is normal and, providing the exhaust smoke cleans up once the engine is at running temperature, nothing should be done. Technician B says that there could be a problem with the cold-start strategy and the engine should be immediately checked. Who is correct?
    A. A only
    B. B only
    C. Both A and B
    D. Neither A nor B   (A5)

19. Camshaft bearing diameter should be measured with a
    A. feeler gauge.
    B. inside micrometer.
    C. Plastigage.
    D. Manometer.   (C6)

20. Technician A says that only manufacturer software is compatible with diagnostic data bus protocols. Technician B says that each manufacturer uses a different data link connector, so only their diagnostic instruments can be used. Who is correct?
    A. A only
    B. B only
    C. Both A and B
    D. Neither A nor B   (F2.1)

21. All of the following would be used while performing an alternator amperage output test EXCEPT:
    A. a carbon pile load tester
    B. amp clamp
    C. refractometer
    D. voltmeter   (G7)

22. Technician A says before installation, head bolts must be cleaned and inspected for erosion or pitting. Technician B says the cylinder deck needs to be checked for warpage before installing the head. Who is correct?
    A. A only
    B. B only
    C. Both A and B
    D. Neither A nor B   (B2)

23. Technician A says that it is permitted to mix ethylene glycol (EG) and propylene glycol (PG) antifreeze in a cooling system but that the mixture concentration can only be tested using a refractometer. Technician B says that extended life coolant mixture should never have water added to it. Who is correct?
    A. A only
    B. B only
    C. Both A and B
    D. Neither A nor B   (D10)

24. Which of the following would be Most-Likely to cause a fuel starvation problem severe enough to shut down an engine?
    A. plugged fuel tank breather
    B. intake manifold leak
    C. insufficient throttle arm travel
    D. missing fuel tank cap   (A12)

25. Technician A says most oil pump pressure-regulating valves are factory-sealed devices. Technician B says if an oil pump pressure-regulating valve fails, the oil pump must be replaced. Who is correct?
    A. A only
    B. B only
    C. Both A and B
    D. Neither A nor B   (D3)

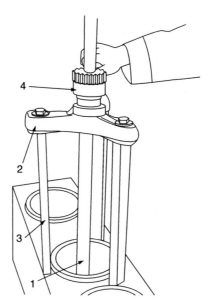

26. Technician A says the tool as shown in the figure is used to bore out cylinder liners. Technician B says it is to check liner protrusion. Who is correct?
    A. A only
    B. B only
    C. Both A and B
    D. Neither A nor B   (C5)

27. Technician A says that most engine management computers permit fault codes to read by flashing codes through dash diagnostic lights. Technician B says that digital driver displays (DDD) on the dash on many current vehicles will alert the operator to any conditions that could affect the operation of the vehicle. Who is correct?
    A. A only
    B. B only
    C. Both A and B
    D. Neither A nor B   (F2.1)

28. An engine being overhauled is found to have pistons with cracked skirts. Technician A says this could have been caused by excessive piston-to-cylinder wall clearance. Technician B says this could have been caused by running the engine with a low oil level. Who is correct?
    A. A only
    B. B only
    C. Both A and B
    D. Neither A nor B   (C12)

29. Droplets of water or gray oil on the dipstick would LEAST-Likely be caused by a
    A. cracked block or cylinder head.
    B. blown head gasket.
    C. leaking oil cooler bundle.
    D. worn oil pump.   (A2)

30. Technician A says that low manifold boost pressure can be caused by air inlet restriction. Technician B says that you can check turbo boost pressure with a water manometer. Who is correct?
    A. A only
    B. B only
    C. Both A and B
    D. Neither A nor B   (A8)

31. Technician A says valve rotators can be tested using a plastic mallet tapping the valve stems. Technician B says defective valve seals will cause oil consumption. Who is correct?
    A. A only
    B. B only
    C. Both A and B
    D. Neither A nor B   (B4)

32. A vehicle is checked for poor acceleration. The engine idles normally, but is slow to generate boost pressure. Once the vehicle is at higher road speeds, the vehicle runs normally. Which could be the Most-Likely cause of this condition?
    A. an inoperative wastegate actuator
    B. a leaking intake manifold gasket
    C. a clogged air filter
    D. a plugged EGR port   (E3)

33. To maintain an effective cooling system, all of the following must be done EXCEPT:
    A. The radiator and engine must be back-flushed annually.
    B. Coolant should be tested for SCA strength and acidity.
    C. Air must be bled from the engine during refilling of the system.
    D. Coolant should be tested for antifreeze protection level.   (A14)

34. Technician A says that an easy way to erase fault codes on most diesel engine ECMs is to disconnect the vehicle batteries. Technician B says that when the vehicle batteries are disconnected, none of the tattletales and audit trails are lost because they are written to EEPROM. Who is correct?
    A. A only
    B. B only
    C. Both A and B
    D. Neither A nor B   (A16)

35. Technician A says that some ELC coolants require only one SCA treatment in six years. Technician B says that engine cooling fan engagement may be included in engine braking strategy. Who is correct?
    A. A only
    B. B only
    C. Both A and B
    D. Neither A nor B   (A14)

36. You suspect a cylinder head has a blown head gasket. The cylinder head is still installed on the engine. What would you do first?
    A. Check head bolt torque.
    B. Magnetic flux test the cylinder head.
    C. Remove all residue oil, carbon, and gasket material and visually inspect for cracks.
    D. Perform a pressure test of the head to check for internal cracks.   (B3)

37. Technician A says that extended life coolants (ELCs) are premixed and should not be diluted with water. Technician B says that silicone radiator hoses should have special clamps. Who is correct?
    A. A only
    B. B only
    C. Both A and B
    D. Neither A nor B   (D10)

38. Technician A says cylinder heads are always torqued in sequence according to specifications set by the manufacturer. Technician B says that any pushrods that are bent must be straightened. Who is correct?
    A. A only
    B. B only
    C. Both A and B
    D. Neither A nor B   (B7)

39. An engine produces a deep-sounding thumping noise that is especially noticeable at slow rpm and high engine loads. Technician A says this could be caused by worn crankshaft main bearings. Technician B says that this could be a bottom-end knock caused by a failed rod big-end journal. Who is correct?
    A. A only
    B. B only
    C. Both A and B
    D. Neither A nor B   (A4)

40. A bus enters the repair shop with damaged engine door wiring for the taillights. Technician A says repair the wiring and place the vehicle back in service. Technician B says missing engine door stops may have caused the wiring damage. Who is correct?
    A. A only
    B. B only
    C. Both A and B
    D. Neither A or B   (A17)

41. A vehicle with mechanical fuel controls is being checked for a sticking throttle complaint. Technician A says the lever return springs may be weak. Technician B says there may be air in the fuel lines. Who is correct?
    A. A only
    B. B only
    C. Both A and B
    D. Neither A nor B   (F1.5)

42. Technician A says that the customer data programming can be changed in electronic systems with a PC and proprietary software. Technician B says in some cases changing data programming could result in *less* efficient engine performance. Who is correct?
    A. A only
    B. B only
    C. Both A and B
    D. Neither A nor B   (F2.8)

43. Technician A says to use a straightedge across the exhaust manifold surfaces to align separate cylinder heads on the engine block. Technician B says to blow out head bolt holes to ensure bolts do not bottom-out prematurely. Who is correct?
    A. A only
    B. B only
    C. Both A and B
    D. Neither A nor B   (B2)

44. Technician A says by using a pressure relief valve in the surge tank, the boiling point of the coolant is raised. Technician B says that antifreeze raises the boil point of water. Who is correct?
    A. A only
    B. B only
    C. Both A and B
    D. Neither A nor B   (D8)

45. Technician A says a testing instrument panel oil pressure gauge should closely follow a master gauge. Technician B says test readings should be taken and compared during startup, varying operating ranges, and on shutdown. Who is correct?
    A. A only
    B. B only
    C. Both A and B
    D. Neither A nor B   (A15)

46. The piston shown above came from an in-line 6-cylinder diesel engine equipped with electronic unit injection. All the other pistons showed normal wear. This was the only piston showing damage. Which of the following is the Most-Likely cause?
    A. worn exhaust cam lobe
    B. restricted exhaust system
    C. restricted air cleaner
    D. worn injector cam lobe   (C15)

47. Technician A says before installing die-cast rocker covers on a cylinder head, ensure a silicone gasket is coated with adhesive. Technician B says to always use new gaskets when installing valve covers. Who is correct?
    A. A only
    B. B only
    C. Both A and B
    D. Neither A nor B   (C1)

48. On a vehicle with an electric fuel shut-off valve, which of the following would not be part of the circuit?
    A. fuel pressure sensor
    B. key switch
    C. solenoid armature
    D. solenoid coil   (F1.2)

49. A mechanically controlled engine will crank but will not start. If ether is used, the engine will run for a few seconds, and then die. What could be the Most-Likely cause?
    A. a clogged air filter
    B. a defective fuel shut-off solenoid
    C. low engine oil level
    D. a bad starter   (F1.11)

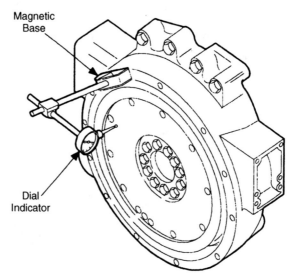

50. What is being performed in the figure?
    A. clutch face surface runout
    B. pilot bore radial runout
    C. crankshaft radial runout
    D. flywheel housing concentricity   (C18)

51. An EGR-equipped vehicle is being checked for overheating. Technician A says the EGR cooler may be clogged, causing the engine to overheat. Technician B says the EGR valve may be leaking, allowing EGR flow all the time. Who is correct?
    A. A only
    B. B only
    C. Both A and B
    D. Neither A nor B   (E7)

52. Which of the following would LEAST-Likely cause excessive backlash in the timing gear train?
    A. worn idler gear bearings
    B. worn gear teeth
    C. worn camshaft bearings
    D. incorrect timing mark alignment   (C10)

53. What should be used to measure ring-to-groove clearance?
    A. Plastigage
    B. micrometer
    C. feeler gauges
    D. dial bore gauge   (C13)

54. All of the following are used to determine battery performance and condition EXCEPT:
    A. multimeter
    B. carbon pile load tester
    C. refractometer
    D. voltage drop from battery post to positive terminal   (G2)

55. Technician A says that oil in a head bolt hole in the block can throw off the torque setting of that bolt. Technician B says that a head bolt with stripped threads should be dressed before installing. Who is correct?
    A. A only
    B. B only
    C. Both A and B
    D. Neither A nor B   (C2)

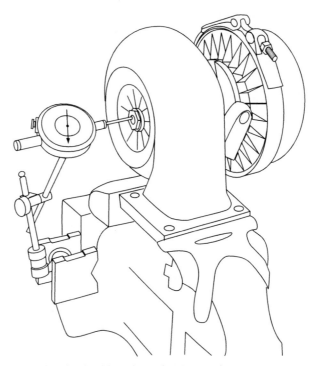

56. In the figure, the technician is checking the turbocharger for
    A. radial runout.
    B. turbine shaft endplay.
    C. turbine fin wear.
    D. compressor fin wear.   (E2)

57. An engine equipped with a **pump-line-nozzle** (PLN) fuel system, cranks, but fails to start and the operation of the primer pump fails to build pressure. Which of the following would MOST-Likely cause this condition?
    A. injector nozzle valve seized
    B. primary fuel filter housing leak
    C. restriction in the high-pressure line to the injector
    D. lines between the prime pump and the injection pump restricted   (F1.3)

58. Technician A says that when performing a starter load test, you should not crank the engine for more than 30 seconds at a time. Technician B says you must perform this test with a battery charger connected to the battery. Who is correct?
    A. A only
    B. B only
    C. Both A and B
    D. Neither A nor B   (G8)

59. An air intake is being checked for a restriction using a water manometer. Technician A says that if the air restriction is excessive, the filter may need to be inspected. Technician B says that the engine speed should be at full load and full rpm to take the readings. Who is correct?
    A. A only
    B. B only
    C. Both A and B
    D. Neither A nor B   (A7)

60. If an injector sleeve (tube) is found to be leaking, which of following procedures is LEAST-Likely to be required?
    A. replacement of the sleeve
    B. performing a cylinder head pressure test
    C. checking nozzle-tip protrusion
    D. replacing the injector nozzle   (B5)

61. The figure shows what type of fan hub?
    A. air clutch fan hub
    B. electrically activated fan hub
    C. thermatic fan hub
    D. hydraulic fan hub   (D14)

62. An engine with a distributor-type injector pump cranks, but fails to start. Which of the following could be a possible cause?
    A. a defective injector
    B. a defective starter cut-out relay
    C. a defective shutdown solenoid
    D. a leaking air intake hose   (F1.6)

63. Technician A says a radiator cap in a cooling system that cannot maintain its rating pressure can cause an overheat condition. Technician B says when the pressure exceeds its rating, the cap relieves the pressure. Who is correct?
    A. A only
    B. B only
    C. Both A and B
    D. Neither A nor B   (D13)

64. On a diesel engine with a charge-air cooler, a technician notices soot in the intake manifold. What should the technician check first?
    A. intercooler
    B. turbocharger
    C. intake manifold
    D. head gasket   (E2)

65. After the technician cleans a disassembled engine block, what should be done next?
    A. paint the exterior
    B. magnetic flux test for cracks
    C. remove the cylinder sleeves
    D. plug all oil ports   (C2)

66. When the oil filter by-pass valve opens, what has caused it to do so?
    A. restricted inlet to the filter
    B. the oil pump is sucking air
    C. massive oil leak in the lubrication system
    D. filter element is plugged   (D3)

67. The rear two cylinders on an in-line 6-cylinder engine show eroded piston wear. Technician A says this could be caused by low water pump flow. Technician B says this could be caused by dirty oil. Who is correct?
    A. A only
    B. B only
    C. Both A and B
    D. Neither A nor B   (D12)

68. Which of the following is LEAST-Likely to produce a loss of power in a diesel at full-rated loads?
    A. high exhaust back pressure
    B. high manifold boost pressure
    C. oil and dirt in the air tubes or cooling fins of the aftercooler
    D. small leak in the fuel line from the tank to the pump   (A12)

69. Technician A says that a worn diesel engine lube pump may cause low oil pressure. Technician B says that the high-pressure oil pump used on HEUI systems uses a swash plate principle. Who is correct?
    A. A only
    B. B only
    C. Both A and B
    D. Neither A nor B   (D2)

70. Technician A says that main bearing shells should receive a light coat of oil on both the face and back before final installation. Technician B says that when checking the bearing, install Plastigage in all the bearings, torque to specifications, and rotate the crankshaft. Who is correct?
    A. A only
    B. B only
    C. Both A and B
    D. Neither A nor B   (C9)

71. Technician A says that high-pressure fuel lines should be bled of any air that might be trapped after replacement. Technician B says you can check for air in the fuel subsystem by installing a diagnostic sight glass and look for bubbles. Who is correct?
    A. A only
    B. B only
    C. Both A and B
    D. Neither A nor B   (F1.3)

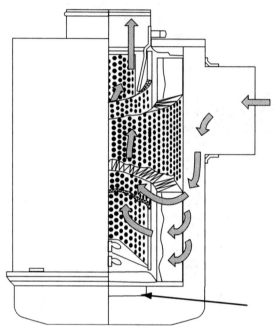

72. What type of air cleaner is shown in the figure?
    A. oil bath
    B. dry type
    C. centrifugal
    D. precleaner   (E1)

73. Technician A says the cylinder head should be checked for warpage with a dial indicator. Technician B says that the wrist pins should be checked for wear with a dial indicator. Who is correct?
    A. A only
    B. B only
    C. Both A and B
    D. Neither A nor B   (B3)

74. While discussing a low voltage problem at the starter, Technician A says that a corroded battery cable may be the cause. Technician B says that a defective starter drive may be the cause. Who is correct?
    A. A only
    B. B only
    C. Both A and B
    D. Neither A nor B   (G5)

75. A vehicle is found to have high exhaust back pressure and elevated exhaust temperature. Technician A says this is caused by a DPF that needs to be serviced. Technician B says that it is caused by a burned exhaust valve. Who is correct?
    A. A only
    B. B only
    C. Both A and B
    D. Neither A nor B   (A9)

76. When rebuilding an engine, Technician A says the viscous crankshaft damper should always be replaced. Technician B says cracks in the damper can be welded and the damper reused. Who is correct?
    A. A only
    B. B only
    C. Both A and B
    D. Neither A nor B   (C16)

77. An electronically controlled diesel engine runs poorly. It has 8 active codes and 10 inactive codes. All the codes were set at the same ECM time. Which of the following is the most likely cause?
    A. worn crankshaft thrust washer
    B. worn camshaft thrust washer
    C. poor electrical supply connections
    D. faulty ECM   (F2.9)

78. A camshaft fails to pass visual inspection. Technician A says damaged camshaft bearing surfaces can be turned. Technician B says flat spots on lobes can be silver soldered and machined to original specifications. Who is correct?
    A. A only
    B. B only
    C. Both A and B
    D. Neither A nor B   (B11)

79. Before installing the valves into the cylinder head it is important to check which one of the following?
    A. valve-to-seat contact
    B. rocker arm adjustment
    C. camshaft lift
    D. head gasket   (B7)

80. When installing a camshaft into the cylinder block, the technician finds it is hard to turn. Which of these could be the LEAST-Likely cause?
    A. the camshaft is bent
    B. the camshaft bore is out of line
    C. bearing-to-journal clearance is incorrect
    D. the camshaft gear was not installed prior to camshaft installation   (C7)

81. An electronically controlled diesel engine will start but will only idle. Technician A says a faulty intake air temperature sensor could cause this. Technician B says a faulty TPS could cause this. Who is correct?
    A. A only
    B. B only
    C. Both A and B
    D. Neither A nor B   (F2.2)

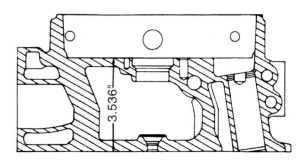

82. The cylinder head deck-to-deck thickness measurement indicated in the figure is used to
    A. prevent cylinder head warpage.
    B. calculate valve guide height.
    C. determine cylinder warpage.
    D. determine resurfacing limit.   (B3)

83. Which is LEAST-Likely to be used to check connections on the electronic circuits?
    A. digital ohmmeter
    B. digital multimeter
    C. test light
    D. breakout box   (F2.6)

84. In the customer data programming options in the software of an Electronic Service Tool (EST), the technician can select between LS and VS governing. Which option would make the vehicle accelerator pedal respond similarly to the accelerator pedal in an automobile?
    A. LS
    B. VS
    C. no difference
    D. both options would have to be selected   (F2.8)

85. A bus driver requests that logged incidents of engine overspeed and panic braking be removed from the engine ECM audit trails. Technician A says that he can do this by erasing all logged data using a PC and proprietary software. Technician B says that this can also be done by replacing the ECM PROM chips. Who is correct?
    A. A only
    B. B only
    C. Both A and B
    D. Neither A nor B   (F2.1)

86. A camshaft fails to pass visual inspection. Technician A says you cannot machine defective camshaft bearing surfaces. Technician B says you visually inspect the camshaft gear for cracks, chips, and broken teeth. Who is correct?
    A. A only
    B. B only
    C. Both A and B
    D. Neither A nor B   (B11)

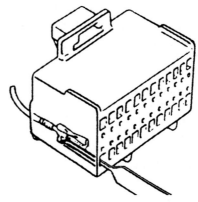

87. What is being done in the figure?
    A. A voltage drop test.
    B. The locking tang is being released.
    C. The terminal blades are being spread.
    D. A digital multimeter test probe is being inserted.   (A3)

88. All of the following must be checked before installing rocker arms, EXCEPT:
    A. rocker arm bushings for wear
    B. rocker arm shaft for wear
    C. rocker arm pallet for pitting or scoring
    D. rocker arm adjustment screws are down   (B8)

89. Technician A says that a typical throttle position sensor is connected to a 5-volt reference. Technician B says when the accelerator is depressed, signal voltage will rise above 5 volts. Who is correct?
    A. A only
    B. B only
    C. Both A and B
    D. Neither A nor B   (F2.2)

90. Technician A says that temperature sensing devices used on electronically managed diesel engines are thermistors, which receive a reference voltage (V-Ref) and return a portion of it as a signal to the ECM. Technician B says that some vehicle thermistors use a negative temperature coefficient (NTC), meaning that resistance decreases as temperature goes up. Who is correct?
    A. A only
    B. B only
    C. Both A and B
    D. Neither A nor B   (F2.9)

91. Which of these would be LEAST-Likely to cause low engine oil pressure?
    A. worn bearings
    B. use of oil with a high viscosity index
    C. restricted suction tube
    D. worn oil pump   (A15)

92. Technician A says that a gray or black smoke from the engine exhaust could be an indication of air starvation. Technician B says that gray or black smoke can be caused by overfueling. Who is correct?
    A. A only
    B. B only
    C. Both A and B
    D. Neither A nor B   (A5)

93. When replacing a coolant filter, Technician A says to prime the filter with coolant before installation. Technician B says to turn the service valve off before removing the old filter. Who is correct?
    A. A only
    B. B only
    C. Both A and B
    D. Neither A nor B   (D11)

94. In a 2 12-V battery, 12-V electrical system, how are the batteries connected?
    A. series circuit
    B. parallel circuit
    C. series/parallel circuit
    D. isolation circuit   (G8)

95. A crankshaft is found to have fractures that may have been caused by improper alignment. Technician A says to use a master bar to check the main bore alignment. Technician B says to use an inside micrometer to check the main bore diameter for an out-of-round condition. Who is correct?
    A. A only
    B. B only
    C. Both A and B
    D. Neither A nor B   (C8)

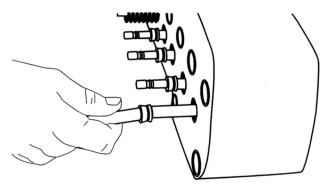

96. What procedure is the technician performing on the cylinder head in the figure?
    A. replacing valve guide seals
    B. replacing valve guides
    C. replacing valve stems
    D. replacing rocker arm studs   (B4)

97. Technician A says that when replacing solid lifters you should also replace the timing gears. Technician B says that each lifter should be checked in its mating bore in the block to ensure proper fit. Who is correct?
    A. A only
    B. B only
    C. Both A and B
    D. Neither A nor B   (B9)

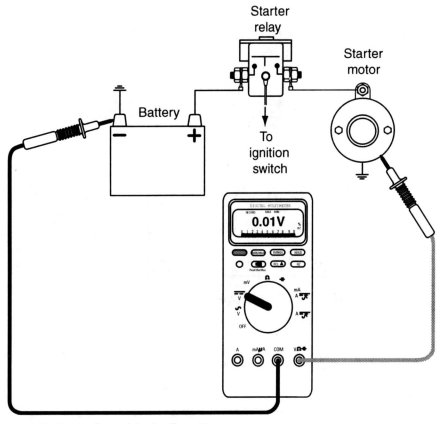

98. What test is being performed in the figure?
    A. starter motor source voltage
    B. voltage drop of the starter ground circuit
    C. voltage drop of the starter positive circuit
    D. resistance in the starter control circuit   (G8)

99. When installing head bolts, Technician A says a service manual is the best place to find head
    bolt torque procedures. Technician B says to check for factory service bulletins before
    proceeding with any procedure. Who is correct?
    A. A only
    B. B only
    C. Both A and B
    D. Neither A or B   (A18)

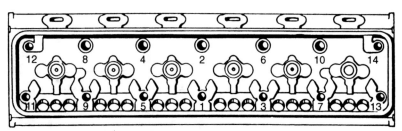

6-Cylinder Engine Cylinder Head

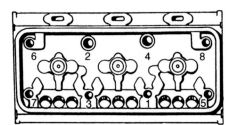

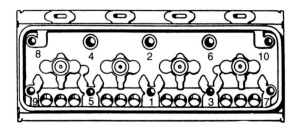

3-Cylinder Engine Cylinder Head   4-Cylinder Engine Cylinder Head

100. The figure shows a sequence of numbers. These numbers are used for what purpose?
    A. to tell the technician which bolts to put in first
    B. to identify the torque sequence
    C. the serial number of the head
    D. the part number of the head   (B7)

101. The throttle position sensor is being adjusted on an engine with electronic controls. Technician A says that on some applications the throttle position sensor produces a PWM (pulse width modulated) return signal. Technician B says the software will indicate when the setting is too low or too high. Who is correct?
    A. A only
    B. B only
    C. Both A and B
    D. Neither A nor B   (F2.9)

102. Technician A says that testing the oil cooler bundle should be done at PM intervals. Technician B says that testing the oil cooler bundle is done as part of an engine overhaul. Who is correct?
    A. A only
    B. B only
    C. Both A and B
    D. Neither A nor B   (D4)

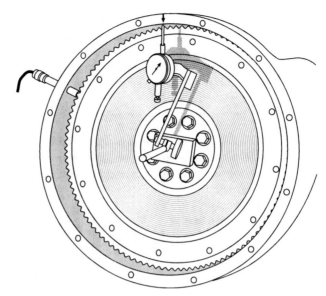

103. In the setup shown in the figure, what is being checked on the flywheel housing?
    A. lateral runout
    B. face lateral runout
    C. radial runout
    D. pilot bore radial runout   (C18)

104. Technician A says "lugging" occurs when the accelerator is in full fuel position and the engine cannot pick up speed under load. Technician B says air starvation results in low cylinder temperatures. Who is correct?
    A. A only
    B. B only
    C. Both A and B
    D. Neither A nor B   (A1)

105. A fuel pump with a mechanical variable speed governor does all of the following EXCEPT:
    A. controls idle speed
    B. meters fuel to the injectors
    C. regulates maximum engine speed
    D. prevents engine lug   (F1.6)

106. Technician A says dash gauges provide better information than warning lights because you can see when they are not working. Technician B says failure of a digital or electronic gauge may not be the fault of the actual gauge circuit. Who is correct?
    A. A only
    B. B only
    C. Both A and B
    D. Neither A nor B   (D1)

107. Technician A says that in a HEUI fuel system, high-pressure engine oil is used as hydraulic medium to actuate the injector fuel pump stroke. Technician B says that because the engine ECM controls the actuation oil pressure in a HEUI system, the rate of injecting fuel can be precisely controlled. Who is correct?
    A. A only
    B. B only
    C. Both A and B
    D. Neither A nor B   (F2.3)

108. When performing a load test on a fully charged battery, the technician finds that the battery voltage drops below manufacturer's specifications. The technician should do which one of the following?
    A. recharge the battery and return it to service
    B. recharge the battery and retest it
    C. replace the battery
    D. replace the voltage regulator   (G2)

109. While discussing a computer-controlled diesel engine, Technician A says corrosion on wiring connectors can affect engine performance. Technician B says that after repairing sealed wire connectors, they should be treated with a corrosion inhibitor. Who is correct?
    A. A only
    B. B only
    C. Both A and B
    D. Neither A nor B   (F2.6)

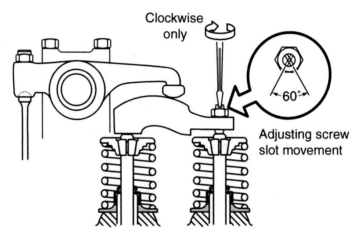

110. In the figure, what adjustment is being performed?
    A. valve bridge/yoke balance
    B. valve lash
    C. injector timing
    D. injector hold-down   (B6)

111. Technician A says a micrometer can be used to measure the crankshaft journals. Technician B says that when cleaning crankshaft journals, a wire wheel on a die grinder can be used. Who is correct?
    A. A only
    B. B only
    C. Both A and B
    D. Neither A nor B   (C8)

112. Technician A says the starter ring gear should be replaced if damaged teeth are found. Technician B says the flexplate bolts directly to the crankshaft. Who is correct?
    A. A only
    B. B only
    C. Both A and B
    D. Neither A nor B   (C18)

113. Technician A says ethylene glycol (EG) and propylene glycol (PG) should not be mixed. Technician B says if switching the type of antifreeze used, the cooling system must be flushed completely first. Who is correct?
     A. A only
     B. B only
     C. Both A and B
     D. Neither A nor B   (D10)

114. Which of the following must the technician do first after dry cylinder liners are removed?
     A. main bearing caps must be inspected
     B. lock bores must be inspected
     C. oil ports must be inspected
     D. cam bearing bores must be inspected   (C4)

115. A vehicle has been found to have excessive dirt in the fuel filters. Technician A says to change the filters and return the vehicle to service. Technician B says in addition to servicing the filters, the fuel tank should be drained and flushed. Who is correct?
     A. A only
     B. B only
     C. Both A and B
     D. Neither A or B   (A6)

116. Technician A says that electronic connectors should never be probed through the weather-pack seals. Technician B says a breakout box or T should be used to test connectors. Who is correct?
     A. A only
     B. B only
     C. Both A and B
     D. Neither A nor B   (F2.6)

117. Which of the following should the technician check first when installing a cylinder head to the engine block?
     A. bolt holes for foreign matter and cross threading
     B. head bolt torque
     C. rocker pedestal height
     D. valve bridge adjustment   (B7)

118. A 4-stroke cycle diesel engine has a vibration occurrence rate that is equal to half the engine rpm. Technician A says that this might be caused by a bent camshaft. Technician B says that this could be caused by a bent connecting rod. Who is correct?
     A. A only
     B. B only
     C. Both A and B
     D. Neither A nor B   (A13)

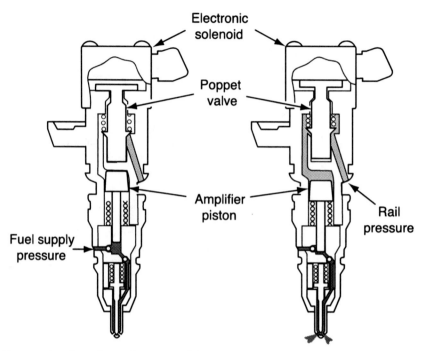

119. In the figure, what force is used to create injection pressure values?
    A. hydraulic
    B. electronic
    C. electromechanical
    D. mechanical   (F2.3)

120. A diesel ECM has logged both active and inactive codes. Technician A says that both the active
    and inactive codes can be erased providing the manufacturer's diagnostic software is used.
    Technician B says that both active and inactive codes produce fault mode indicators (FMIs),
    which permit them to be accessed by any diagnostic instrument capable of reading SAE codes.
    Who is correct?
    A. A only
    B. B only
    C. Both A and B
    D. Neither A nor B   (F2.1)

121. Technician A says if the turbocharger lines and hoses become frayed or nicked, they must be
    replaced not repaired. Technician B says turbocharger oil tubes must be checked for
    obstructions. Who is correct?
    A. A only
    B. B only
    C. Both A and B
    D. Neither A nor B   (E2)

122. Technician A says overfueling can cause black smoke emission. Technician B says that coolant
    in the combustion chamber can cause this problem. Who is correct?
    A. A only
    B. B only
    C. Both A and B
    D. Neither A nor B   (A5)

123. A vehicle is equipped with a serial bus dash. All gauges read erratic. The ECM has not turned on the check engine lamp. Which of the following is the Most-Likely cause?
A. faulty dash
B. faulty alternator
C. a faulty oil pressure sensor
D. a faulty oil pressure gauge   (D1)

124. What should the battery's electrolyte temperature not be allowed to exceed during the "fast charge" method?
A. 125°F
B. 200°F
C. 100°F
D. 150°F   (G3)

125. While inspecting an electrical connector for damage, the connector lock tab is found to be broken off. Technician A says to use a wire-tie to secure the connector. Technician B says to replace the connector and install new terminal pins. Who is correct?
A. A only
B. B only
C. Both A and B
D. Neither A or B   (F2.7)

126. When installing the wrist pin retaining snap rings in an articulating piston assembly, what is most commonly the recommended position of the snap ring gap?
A. downward
B. upward
C. horizontal
D. does not matter   (C11)

127. A vehicle has higher than normal fuel pressure. Which could be the Most-Likely cause?
A. a broken regulator valve spring
B. a blocked fuel tank vent
C. a stuck regulator valve piston
D. aerated fuel   (F1.4)

128. A vehicle is being diagnosed for white smoke on cold weather startups. When the master switch is turned on, the "Wait To Start" lamp illuminates and goes out, but when the engine is started it emits white smoke. Technician A says the heater grids should be checked with an ohmmeter. Technician B says the heaters are controlled by relays, and the relays should be replaced. Who is correct?
A. A only
B. B only
C. Both A and B
D. Neither A nor B   (E6)

129. While inspecting the exhaust system, the technician notices that the vehicle has excess exhaust back pressure. Which of the following is LEAST-Likely to cause this?
A. improper muffler
B. carbon build-up in the exhaust system
C. collapsed exhaust pipe
D. high exhaust temperature   (A9)

130. When testing a cylinder head for potential coolant leakage, Technician A says the coolant needs to be heated to operating temperature. Technician B says the coolant needs to be under pressure. Who is correct?
    A. A only
    B. B only
    C. Both A and B
    D. Neither A nor B   (B3)

131. A connector is found to be distorted from heat but when checked with an ohmmeter the electrical connection is still good. Technician A says to replace the connector. Technician B says to apply electrical tape around the connector. Who is correct?
    A. A only
    B. B only
    C. Both A and B
    D. Neither A nor B   (F2.10)

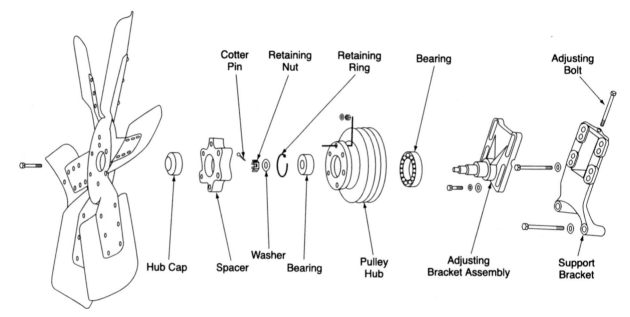

132. In the fan assembly shown in the figure, Technician A says that fan bearings should be submerged in cleaning solvent and air-dried. Technician B says because fan bearings are water-cooled some corrosion is acceptable, although if corrosion is excessive they should be replaced. Who is correct?
    A. A only
    B. B only
    C. Both A and B
    D. Neither A nor B   (D14)

133. Which of the following is LEAST-Likely to be the cause of a electronic unit injector (EUI) diesel engine not starting?
    A. excessive exhaust back pressure
    B. short in the starter armature
    C. corrosion on battery terminals
    D. low engine lube level   (A11)

134. An engine is being overhauled because of a broken crankshaft. Technician A says the operator could have been overspeeding the engine. Technician B says an out-of-round condition may exist on one or more of the main bores. Who is correct?
  A. A only
  B. B only
  C. Both A and B
  D. Neither A nor B   (A1)

135. Technician A says that with a wet sleeve-type cylinder, the cylinder only seals at the top. Technician B says in this type of liner the coolant comes into direct contact with the liner. Who is correct?
  A. A only
  B. B only
  C. Both A and B
  D. Neither A nor B   (C3)

136. Technician A says that an oil residue on the inside of the engine door may indicate excess crankcase pressure. Technician B says that oil on the exhaust joints could indicate a turbocharger oil seal failure. Who is correct?
  A. A only
  B. B only
  C. Both A and B
  D. Neither A nor B   (A2)

137. When removing an internal shutdown solenoid from a rotary distributor pump, what must the technician remove first?
  A. the solenoid cover
  B. the pump cover
  C. the injector pump
  D. the control module   (F1.6 and F2.5)

138. All of the following are components in the valvetrain EXCEPT:
  A. rocker
  B. push tube
  C. cam follower
  D. cam plug   (B1)

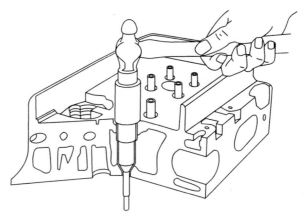

139. After replacing the injector sleeve as shown in the figure, Technician A says the injector tip height must be checked. Technician B says to perform a cylinder head leakage test. Who is correct?
    A. A only
    B. B only
    C. Both A and B
    D. Neither A nor B   (B5)

140. A diesel engine that has a low power problem could be caused by all of the following EXCEPT:
    A. restricted air filter
    B. leaking fuel line
    C. restricted fuel filter
    D. missing air filter   (A12)

141. Technician A says that valve crosshead guide pins should be at right angles to the cylinder head milled surface. Technician B says to check the crosshead guide pin diameter with a micrometer, and compare it to manufacturer's specifications. Who is correct?
    A. A only
    B. B only
    C. Both A and B
    D. Neither A nor B   (B6)

142. An engine fails to achieve normal operating temperature. Technician A says that a stuck thermostat could cause this problem. Technician B says that an overfilled radiator could cause this problem. Who is correct?
    A. A only
    B. B only
    C. Both A and B
    D. Neither A nor B   (A14)

143. Technician A says misalignment between pulleys must not exceed 1/16 in. (1.59 mm) for each 12 in. (30.5 cm) of distance between pulley running centerlines. Technician B says belts must not touch the bottom of grooves or protrude more than 3/32 in. (2.38 mm) above the top of the groove. Who is correct?
    A. A only
    B. B only
    C. Both A and B
    D. Neither A nor B   (D12)

144. When the accelerator pedal on an electric foot pedal assembly (EFPA) is moved from 0 to full travel, the throttle position sensor (TPS) should indicate
    A. 0 VDC to 5 VDC.
    B. 5 VDC to 0 VDC.
    C. 0.5 VDC to 4.5 VDC.
    D. 4.9 VDC to 0.2 VDC.   (F2.2)

145. Which would LEAST-Likely be caused by incorrectly aligning the flywheel housing?
    A. rear oil seal failure
    B. broken flywheel bolts
    C. transmission damage
    D. U-joint failure   (C17)

146. Troubleshooting has indicated that a thermistor-type coolant temperature sensor is not returning a signal voltage. Which of the following should be done first?
    A. Check the V-Ref value.
    B. Consult the manufacturer's service literature.
    C. Simulate V-Ref to check for signal voltage.
    D. Check thermistor internal resistance using an ohmmeter.   (D8)

147. An operator complains the fuel gauge is always reading full, even after driving all day. Technician A says the sending unit in the tank may have failed. Technician B says the sending unit float may have a hole in it, and it is filled with fuel. Who is correct?
    A. A only
    B. B only
    C. Both A and B
    D. Neither A nor B   (F1.1)

148. When removing a starter motor from a vehicle, what should the technician remove first?
    A. starter bolts
    B. starter solenoid
    C. battery ground cable
    D. starter magnetic switch   (G6)

149. Fuel injectors should be timed when
    A. the exhaust valves are open.
    B. the injector follower is fully depressed.
    C. the cylinder is at TDC of a compression stroke.
    D. excessive valvetrain noise is noticed.   (B10)

150. Under load, a diesel engine emits dark black smoke from the exhaust pipe. What should the technician check first?
    A. air cleaner
    B. fuel pump
    C. injector pump
    D. turbocharger   (A5)

151. All of the following could cause fuel to contaminate the engine oil EXCEPT:
    A. cracked fuel gallery in the head
    B. broken piston ring
    C. leaking intake manifold gasket
    D. leaking injector seal   (A15)

152. When removing a piston from an engine cylinder block, which of these should the technician do first?
  A. remove the crankshaft
  B. remove the piston wrist pin
  C. remove the carbon ridge
  D. remove the sleeve seals   (C2)

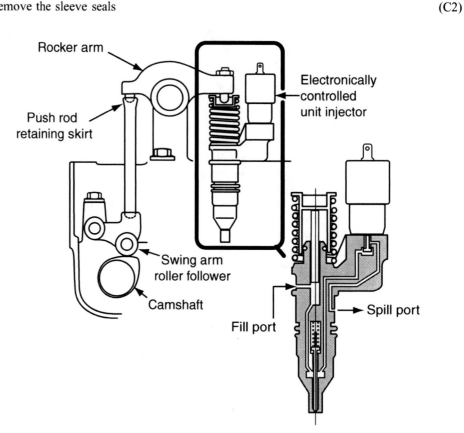

153. The figure shows a sectional view of a typical electronic unit injector (EUI). Technician A says that injection pressures are created by mechanical force from the camshaft. Technician B says that actual injected fuel quantity is computer controlled by switching the control solenoid. Who is correct?
  A. A only
  B. B only
  C. Both A and B
  D. Neither A nor B   (F2.6)

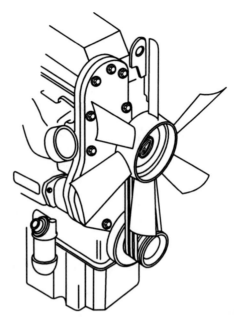

154. A thermostatically controlled fan, as shown in the figure, may rely on all of the following to regulate fan operation EXCEPT:
    A. a coil spring
    B. air temperature
    C. radiator temperature
    D. air pressure   (D14)

155. A fleet bus driver complains that his electronically managed engine will not reach rated engine rpm when in the upper transmission gear ratios. Technician A says that there is a problem with the fuel map programming to the engine ECM, limiting top-end fueling. Technician B says that the engine electronics have been programmed with a progressive shifting profile designed to improve fuel economy. Who is correct?
    A. A only
    B. B only
    C. Both A and B
    D. Neither A nor B   (F2.1)

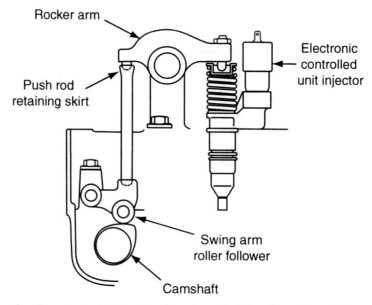

156. In the figure showing a typical EUI, what does the technician do to time the injector when the top end is set?
    A. adjust the rocker arm screw
    B. program the calibration codes to the ECM
    C. program the timing codes to the ECM
    D. adjust the cam follower assembly   (F2.4)

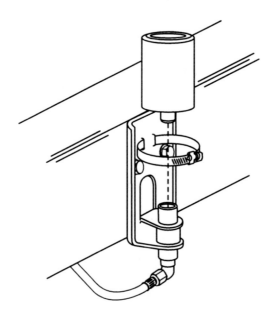

157. An ether injector is shown in the figure. Technician A says that this device is used in conjunction with glow plugs as a cold weather starting aid. Technician B says that excessive use of this device can result in broken piston rings and ring lands. Who is correct?
    A. A only
    B. B only
    C. Both A and B
    D. Neither A nor B   (C11)

Compressor
Housing

Center
Housing

Clamp

Plate

Turbine
Housing

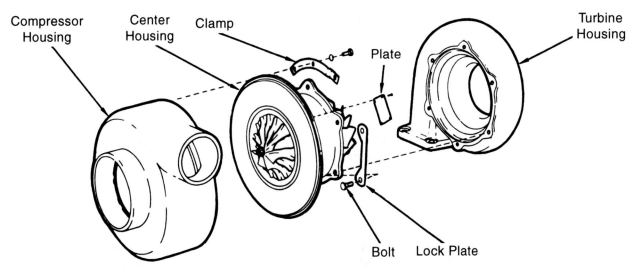

Bolt   Lock Plate

158. Referencing the figure, Technician A says that a turbocharger assembly should never be disassembled because it is impossible to reassemble without expensive balancing instruments. Technician B says that turbochargers are often recored by replacing the center housing as a unit and reusing the old compressor and turbine housings. Who is correct?
    A. A only
    B. B only
    C. Both A and B
    D. Neither A nor B   (E2)

159. Technician A says wet cylinder sleeves should be cleaned by solvent immersion or with glass-bead blasting to the exterior. Technician B says a cylinder out-of-round condition can be detected by measuring at several locations with a dial bore gauge. Who is correct?
    A. A only
    B. B only
    C. Both A and B
    D. Neither A nor B   (C4)

160. The figure shows what part of the lubrication system?
    A. oil filter
    B. oil cooler
    C. filter housing
    D. oil pump   (D4)

161. Technician A says that mechanical rotators on a valve spring will help eliminate carbon deposits on the valve seat. Technician B says that they can prevent oil leakage from the valve guide. Who is correct?
    A. A only
    B. B only
    C. Both A and B
    D. Neither A nor B   (B4)

162. A vehicle has a turbocharger with a seized shaft bearing. Technician A says this could be caused by excessive engine idle time. Technician B says the oil supply line to the turbocharger should be replaced. Who is correct?
    A. A only
    B. B only
    C. Both A and B
    D. Neither A nor B   (D5)

163. Technician A says that a starter solenoid has two coils. Technician B says that the hold-in coil draws more current than the pull-in coil. Who is correct?
    A. A only
    B. B only
    C. Both A and B
    D. Neither A nor B   (G6)

164. Technician A says small coolant leaks can easily be detected, because antifreeze does not readily evaporate. Technician B says a defective thermostat can be caused by repeated overheating of the engine. Who is correct?
    A. A only
    B. B only
    C. Both A and B
    D. Neither A nor B   (A2)

165. A driver complains that a low coolant level alert is intermittently displayed. Technician A says this could be because coolant level in the radiator is low enough so that when the vehicle travels hilly terrain, the coolant level sensor ground is intermittently lost. Technician B says that this condition is normal even if the coolant level is correct, because any sloshing of the coolant in the radiator can trip the low coolant level alert. Who is correct?
    A. A only
    B. B only
    C. Both A and B
    D. Neither A nor B   (F2.1)

166. On a vehicle with an electrical oil pressure gauge all of the following will cause inaccurate readings EXCEPT:
    A. low voltage to the gauge
    B. a short in the wire from the transducer
    C. low oil pressure
    D. a defective transducer   (D1)

167. A recently overhauled engine develops cylinder hydraulic lock. After clearing the liquid from the cylinder, inspecting the heads, and replacing the head gaskets the problem reoccurs. Technician A says the intercooler could be defective. Technician B says the cylinder head may have a hairline crack. Who is correct?
    A. A only
    B. B only
    C. Both A and B
    D. Neither A nor B   (B3)

168. What would the technician do first when adjusting drive belts on an engine?
    A. replace the drive pulleys
    B. replace the idler pulleys
    C. check belts for cracking and wear
    D. apply belt dressing compound   (D7)

169. A vehicle is reported to have poor fuel mileage. Technician A says that missing thermostats could be the cause. Technician B says a leaking fuel line could be the cause. Who is correct?
    A. A only
    B. B only
    C. Both A and B
    D. Neither A nor B   (D9)

170. When installing the starter, which is LEAST-Likely to be done?
    A. connect control wiring to solenoid
    B. align starter motor with flywheel
    C. install starter mounting bolts
    D. replace starter bushings   (G8)

171. A set of dry liners are to be fitted to a diesel engine cylinder block. Technician A says that it is good practice to selectively fit the liners to block by fitting the liner with the largest OD to the block bore with the largest ID and progressing downward. Technician B says that it is important to machine the correct crosshatch pattern into the block bore before fitting each liner. Who is correct?
    A. A only
    B. B only
    C. Both A and B
    D. Neither A nor B   (C3)

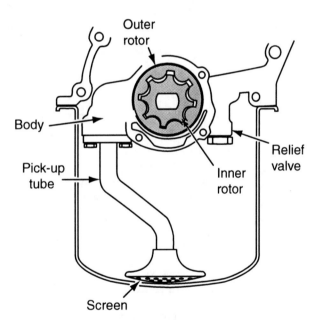

172. What is the function of the component in the figure?
    A. fuel transfer pump
    B. coolant pump
    C. fuel injection pump
    D. oil pump   (D3)

173. Upon inspection of the engine compartment, the engine and most of the engine compartment is covered with a thin layer of black soot. Which of these is the most likely cause?
    A. excessive idling
    B. a malfunctioning boost pressure sensor
    C. a cracked exhaust manifold
    D. loose air intake hose clamps   (E5)

174. Low engine power has been determined to be a result of insufficient fuel delivery to the unit injectors. Which of the following would be the LEAST-Likely cause?
     A. suction circuit leaks in fuel subsystem
     B. defective fuel transfer pump
     C. clogged fuel filter
     D. a leak in the injector return line   (A12)

175. After wet-type cylinder sleeves have been removed, the packing ring area and counterbore show signs of light rust and scale. Technician A says this is normal and no additional action is required except ensuring antifreeze protection includes rust inhibitors. Technician B says to use 100- to 120-grit emery paper to remove rust and scale, and check for erosion and pitting. Who is correct?
     A. A only
     B. B only
     C. Both A and B
     D. Neither A nor B   (C2)

176. Which component of a mechanical unit injector (MUI) system is LEAST-Likely to be the cause of an improper injector spray pattern?
     A. nozzle-tip holes
     B. injection nozzle valve
     C. injector rack
     D. injector follower   (F1.7)

177. Which tool is Most-Likely to be used to determine the starter circuit voltage drop test?
     A. starter shunt resistor
     B. voltmeter
     C. ohmmeter
     D. millimeter   (G8)

178. Which of the following is LEAST-Likely to provide satisfactory results when attempting to identify an engine miss?
     A. Listening to the exhaust to determine which cylinder is misfiring
     B. Using a stethoscope
     C. Manually shorting out each hydromechanical injector
     D. Performing an electronic cylinder cut-out test on computer-controlled engines.   (A4)

179. Which procedure is LEAST-Likely to be used when performing valve seat maintenance?
     A. testing seats for looseness with a ball-peen hammer
     B. using a chisel to remove defective seats
     C. checking seats with a concentricity indicator
     D. dressing a valve seat grinding stone   (B4)

180. When air-pressure testing an engine block, Technician A says the block need only be submerged in hot water for about 20–30 minutes and the water checked for bubbles. Technician B says cracked cylinder blocks should be replaced. Who is correct?
     A. A only
     B. B only
     C. Both A and B
     D. Neither A nor B   (C2)

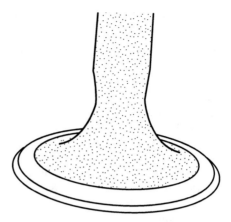

181. The figure shows which of the following valve conditions?
    A. eroded valve stem
    B. carbon deposits on valve face
    C. ridging on valve face
    D. stretched valve stem   (B4)

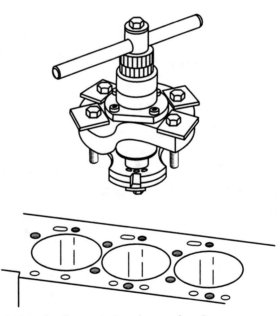

182. For what task is the tool in the figure designed to perform?
    A. remove the liner carbon ridge
    B. remove the cylinder liner
    C. machine the liner crosshatch pattern
    D. cut the liner counterbore   (C3)

183. A vehicle is being checked for an inoperative exhaust brake. Technician A says the vehicle's air
    system pressure should be checked. Technician B says the actuator valve return spring may be
    broken. Who is correct?
    A. A only
    B. B only
    C. Both A and B
    D. Neither A nor B   (E8)

184. When performing failure analysis on a trunk-type piston, heavy erosion and pitting is evident just above the Ni-resist insert on the upper compression ring. Technician A says this is probably caused by fuel with a low CN rating. Technician B says that the cause is advanced fuel injection timing. Who is correct?
    A. A only
    B. B only
    C. Both A and B
    D. Neither A nor B   (C14)

185. Technician A says that when removing high-pressure fuel lines, the seat nipples must be capped to keep out contaminants. Technician B says that overtorquing high-pressure line nuts can cause restricted fuel flow. Who is correct?
    A. A only
    B. B only
    C. Both A and B
    D. Neither A nor B   (F1.10)

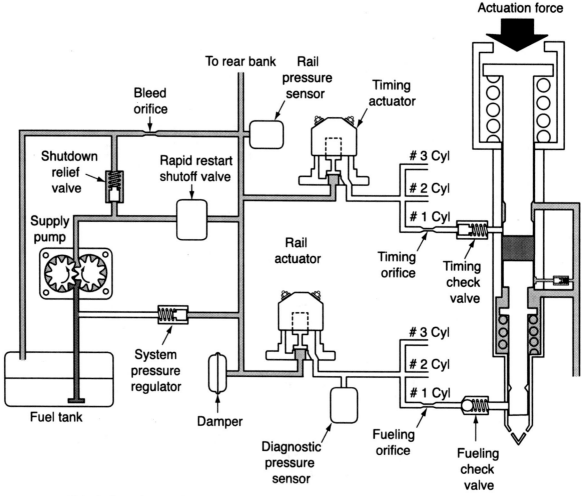

186. Referencing the figure, Technician A says that all of the fuel passing through the timing orifice is injected to the engine during the fuel pulse. Technician B says that fuel to be injected is metered by the rail actuator solenoid. Who is correct?
    A. A only
    B. B only
    C. Both A and B
    D. Neither A nor B   (F2.4)

187. Technician A says to remove the surface charge from a recently charged battery, apply a 300-ampere load across the terminals for 15 seconds. Technician B says that it is unnecessary to remove the surface charge from batteries which have been in storage. Who is correct?
    A. A only
    B. B only
    C. Both A and B
    D. Neither A nor B   (G1)

188. Technician A says the centrifugal oil filter should be serviced at the same time as a regular oil change interval. Technician B says a completely full centrifuge bowl can indicate a severe engine problem. Who is correct?
    A. A only
    B. B only
    C. Both A and B
    D. Neither A nor B   (D6)

189. When installing a wet cylinder sleeve all of the following should be checked EXCEPT:
    A. contact surfaces between the sleeve and the cylinder block
    B. O-ring grooves
    C. bore of the sleeve for scoring
    D. interference fit between the sleeve and block bore   (C5)

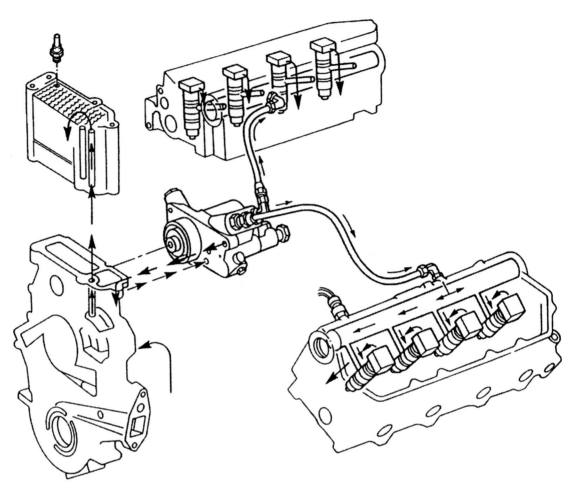

190. The electronically controlled HEUI fuel system shown in the figure has a low power complaint. Technician A says that the pump shown is supplying fuel to the injectors. Technician B says that low actuation oil pressure could be the cause. Who is correct?
   A. A only
   B. B only
   C. Both A and B
   D. Neither A nor B   (F2.3)

191. Technician A says that to test an engine for excess crankcase pressure the engine must be run at low idle. Technician B says the engine must be tested across a range of operating speeds and loads. Who is correct?
   A. A only
   B. B only
   C. Both A and B
   D. Neither A nor B   (A10)

192. A vehicle with a standard 12-V charging/cranking circuit has to be boost started. Technician A says that the battery-to-battery connections should be made with the boost vehicle engine switched off. Technician B says that a portable generator (blast charger) cannot be used on electronically managed engines because high voltage can destroy the engine ECM. Who is correct?
   A. A only
   B. B only
   C. Both A and B
   D. Neither A nor B   (G4)

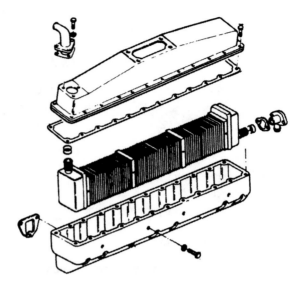

193. Reduced power and high coolant temperature are observed on an engine with an intercooler, as shown in the figure. Technician A says there may be an air lock in the coolant, and it is necessary to bleed air out of the intercooler. Technician B says the air tubes in the cooler may be clogged. Who is correct?
    A. A only
    B. B only
    C. Both A and B
    D. Neither A nor B   (E4)

194. Multiple fractures are discovered in a crankshaft. Technician A says the failure analysis must include inspection of the vibration damper. Technician B says the flywheel should also be closely inspected. Who is correct?
    A. A only
    B. B only
    C. Both A and B
    D. Neither A nor B   (C8)

195. A vehicle equipped with electronically controlled diesel engine has the stop engine light (SEL) illuminated, but not flashing. Technician A says that this indicates the idle shutdown timer has timed out and the engine will shutdown within 60 seconds. Technician B says that the ECM has detected a problem that could result in serious engine damage and, if programmed to, may shutdown the engine. Who is correct?
    A. A only
    B. B only
    C. Both A and B
    D. Neither A nor B   (F2.5)

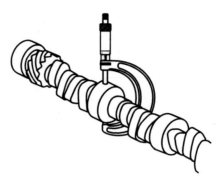

196. What is the technician measuring in the figure?
    A. cam lobe
    B. cam bearing journal diameter
    C. cam lift
    D. cam duration   (B11)

197. Which of the following would more likely result from excessive ether usage during cold startup?
    A. engine runaway
    B. cylinder hydraulic lock
    C. fuel injection system damage
    D. damaged piston compression ring lands   (C11)

198. When replacing the electronic control module (ECM) on an engine with electronic unit injectors (EUIs), what must the technician do first?
    A. disconnect the battery
    B. download customer data
    C. remove injector lines
    D. Turn on the key   (F2.4)

7 Appendices

Answers to the Test Questions for the Sample Test Section 5

1. A	26. B	51. C	76. A
2. A	27. C	52. B	77. A
3. C	28. A	53. C	78. C
4. C	29. C	54. C	79. B
5. B	30. B	55. A	80. C
6. A	31. B	56. A	81. D
7. D	32. B	57. B	82. A
8. D	33. C	58. D	83. B
9. C	34. C	59. B	84. A
10. C	35. A	60. B	85. C
11. C	36. D	61. A	86. C
12. A	37. C	62. B	87. D
13. A	38. D	63. C	88. C
14. C	39. C	64. B	89. A
15. A	40. C	65. A	90. A
16. C	41. B	66. A	91. D
17. C	42. C	67. A	92. B
18. D	43. B	68. A	93. D
19. A	44. A	69. A	94. C
20. A	45. A	70. B	95. B
21. B	46. C	71. A	96. C
22. C	47. A	72. B	97. A
23. B	48. A	73. B	98. B
24. C	49. C	74. C	99. C
25. B	50. A	75. D	

Explanations to the Answers for the Sample Test Section 5

Question #1

Answer A is correct. When filling the fuel tank, the fuel entering displaces the air in the tank. Normally the air escapes through the tank's vent during fueling. If the vent is blocked, the air cannot escape, causing pressure to build. The resulting rise in pressure will cause the nozzle to click off even though the tank is not yet full.

Answer B is incorrect. Water in the fuel tank will have no effect on the operation of the fill nozzle.

Answer C is incorrect. Plugged fuel filters will have no effect on the operation of the fill nozzle.

Answer D is incorrect. The temperature of the fuel will have no effect on the operation of the fill nozzle.

Question #2

Answer A is correct. Only Technician A is correct. Cam profiles are hard surfaced. When the hard surfacing is compromised, the cam is likely to fail. Grooving wear patterns in a camshaft are reason to reject it even if it measures within specification.

Answer B is incorrect. Even if lobe lift measurements are within limits, any grooving or spalling will very quickly turn into a completely ruined camshaft.

Answer C is incorrect. Only Technician A is correct.

Answer D is incorrect. Only Technician A is correct.

Question #3

Answer C is correct. When an engine oil cooler fails, Most-Likely the engine lubricating oil is getting into the cooling system due to its relatively higher pressure. When coolant mixes with oil, it forms a milky gray sludge that tends to float on the coolant and can be seen in the radiator top tank or surge tank.

Answer A is incorrect. A failed oil cooler would not likely affect oil pressure.

Answer B is incorrect. A failed oil cooler would not likely affect oil pressure.

Answer D is incorrect. A leaking oil cooler would most likely cause oil to enter the cooling system. This would not raise the oil level.

Question #4

Answer C is correct. Both Technicians A and B are correct. Valve crosshead guide pins should be checked for squareness and Magnaflux-tested for cracks during an overhaul so both technicians are correct. Valve bridges or yokes push down on two valves at a time. Most engines with four valves per cylinder use valve bridges. They do not usually need to be adjusted during a routine valve adjustment. The bridge must be properly supported to prevent the pedestal shaft from being damaged.

Answer A is incorrect. Technician B is also correct.

Answer B is incorrect. Technician A is also correct.

Answer D is incorrect. Both technicians are correct.

Question #5

Answer B is correct. The steps outlined here to test air intake/inlet restriction are correct with the exception of removing the air cleaner element as the main reason for performing this test to verify the serviceability of the filter element. To perform an air inlet restriction test:

• Connect a vacuum gauge, water manometer, or magnehelic gauge to the intake air piping according to the OEM's instructions.

- Start the engine and operate at the OEM-required load and speed.

- Measure the pressure drop at the turbocharger inlet and record the reading. The typical maximum reading is approximately 25 in. H_2O (635 mm H_2O).

Answer A is incorrect. Connecting a manometer is a step in testing air intake/inlet restriction.

Answer C is incorrect. The engine should be under load.

Answer D is incorrect. The reading is in inches of water.

Question #6

Answer A is correct. Only Technician A is correct. The figure shows an offset camshaft key. The extent of offset determines how the camshaft gear is phased with the camshaft so it can be called a timing key. Technician A is generally correct in saying that the key must be replaced with the same type though if a timing change is required, this is the means of accomplishing it.

Answer B is incorrect. The key shown is not damaged. Actuation of the fuel system pumping pulse by the engine timing gear train must be precisely timed in any diesel engine. This process essentially times the events that manage the fuel system pumping pulse to a specific engine position.

Answer C is incorrect. Only Technician A is correct.

Answer D is incorrect. Only Technician A is correct.

Question #7

Answer D is correct. Open connectors hanging loose in an engine compartment could be signs of tampering or removal of required equipment. The best course of action is to identify the wires and determine where the connectors should be linked. The vehicle or engine manufacturer's service manual can be used to identify wiring codes and colors.

Answer A is incorrect. Loose wiring should be identified before it is secured.

Answer B is incorrect. The wiring should be identified before removing any connectors.

Answer C is incorrect. The use of dummy plugs to protect unused connectors is common practice; however, the technician should first identify the wiring before using dummy plugs.

Question #8

Answer D is correct. The notch in the bearing shell is primarily used to position the bearing shell. Because the notch protrudes into a recess in the bearing bore, it also plays a small role in preventing the bearing from turning. It should be remembered, however, that the primary factor in preventing a friction bearing from turning in its bore is bearing crush. Bearings are retained primarily by crush. The outside diameter of a pair of uninstalled bearing shells slightly exceeds the bore to which it is installed. This creates radial pressure that acts against the bearing halves and provides good heat transfer. The bearing halves are also slightly elliptical to allow the bearing to be held in place during installation; this is known as bearing spread. Tangs in bearings are inserted into notches in the bearing bore to minimize longitudinal movement, prevent bearing rotation, and align oil holes.

Answer A is incorrect. The hole and groove are to direct oil evenly around the bearing.

Answer B is incorrect. The tang is not a defect.

Answer C is incorrect. The stamping on the back indicates an oversized bearing.

Question #9

Answer C is correct. Both Technicians A and B are correct. One method of purging air from the fuel system is by using a hand-primer pump. A fuel system will also have to be purged after the replacement of a fuel hose.

Answer A is incorrect. Technician B is also correct.

Answer B is incorrect. Technician A is also correct.

Answer D is incorrect. Both technicians are correct.

Question #10

Answer C is correct. Both Technicians A and B are correct. They both describe situations that could result in slow cranking.

Answer A is incorrect. Technician B is also correct.

Answer B is incorrect. Technician A is also correct.

Answer D is incorrect. Both technicians are correct.

Question #11

Answer C is correct. Both Technicians A and B are correct. When EUIs are removed from an engine, the fuel rail should first be drained. If it is not, fuel from the fuel manifold can drain into the cylinder. Also, when EUIs are replaced, the new calibration codes must be programmed into the engine management ECM.

Answer A is incorrect. Technician B is also correct.

Answer B is incorrect. Technician A is also correct.

Answer D is incorrect. Both technicians are correct.

Question #12

Answer A is correct. Only Technician A is correct. When an oil pump is disassembled all mating surfaces must be checked to specification.

Answer B is incorrect. The oil pump should be disassembled and thoroughly checked at overhaul. Gerotor pumps tend to wear most between the lobes on the impeller and on the apex of the lobes on the rotor ring. These dimensions should be checked to OEM specifications using a micrometer. The rotor ring-to-body clearance should be checked with a thickness gauge sized to the OEM maximum clearance specification. The axial clearance of the rotor ring and impeller should be measured with a straightedge and thickness gauges.

Answer C is incorrect. Only Technician A is correct.

Answer D is incorrect. Only Technician A is correct.

Question #13

Answer A is correct. An out-of-balance turbocharger will usually result in rapid turbocharger failure and because they are driven by exhaust gas heat (engine rejected heat), rotational speeds do not correlate with engine rpm. Therefore, an out-of-balance turbocharger would be LEAST-Likely to cause an engine speed-related vibration. While driveline vibrations are not uncommon, true engine vibrations are not often a problem. When investigating a vibration complaint, the eliminate all possible causes in the driveline behind the engine first. Remember, the clutch may disengage the driveline from the transmission, but most of the mass of the clutch is rotated with the engine engaged or disengaged.

Answer B is incorrect. A bent connecting rod would cause an engine RPM-related vibration.

Answer C is incorrect. An out-of-balance crankshaft would cause an engine RPM-related vibration.

Answer D is incorrect. A dented vibration damper would render it inoperative, which would create an engine RPM-related vibration.

Question #14

Answer C is correct. Both Technicians A and B are correct. Fast charging does not bring the battery to a fully charged condition suitable for accurate testing results. Any time a battery is charged prior to testing, the surface charge should be removed before testing. Inaccurate test results can occur due to the surface charge.

Answer A is incorrect. Technician B is also correct.

Answer B is incorrect. Technician A is also correct.

Answer D is incorrect. Both technicians are correct.

Question #15
Answer A is correct. The acronym PID is the SAE term for parameter identifier.

Answer B is incorrect. The acronym PID is the SAE term for parameter identifier.

Answer C is incorrect. The acronym PID is the SAE term for parameter identifier.

Answer D is incorrect. The acronym PID is the SAE term for parameter identifier.

Question #16
Answer C is correct. Overconditioned coolant can cause a variety of problem, including sludging, accelerated coolant breakdown, and loss of cooling efficiency. When testing reveals overconditioned coolant, the correct remedy is to drain the cooling system and completely replace the coolant mixture.

Answer A is incorrect. The SCA is already too high; more should not be added.

Answer B is incorrect. If the mixture is extremely overtreated, the bus should be flushed. If the coolant was only slightly overconditioned, this might be a suitable answer.

Answer D is incorrect. The bus should not be run without additives.

Question #17
Answer C is correct. Both Technicians A and B are correct. A particulate filter that needs to be serviced could cause high exhaust back pressure, which would result in low power and high exhaust temperatures. A stuck closed exhaust brake would have the same effect.

Answer A is incorrect. Technician B is also correct.

Answer B is incorrect. Technician A is also correct.

Answer D is incorrect. Both technicians are correct.

Question #18
Answer D is correct. Cylinders 5 and 2 are companioned. Companion cylinders can be identified by pairs in an in-line 6-cylinder engine: cylinders 1 and 6, 5 and 2, and 3 and 4 are companioned; that is, they travel through the cycle together. For instance, when piston #1 is at TDC at the completion of its compression stroke, piston #6 is also at TDC having completed its exhaust stroke. When the crankshaft throws are in line with each other, they are called cylinder throw pairings or companion cylinders. For the firing order 153624, these companion cylinders are 1-6, 5-2, and 3-4. These companion cylinders can be in one of two positions: TDC or TDC overlap. In the case of the 1-6 pair, when #1 piston is at TDC completing its compression stroke, #6 piston, its companion cylinder, is at TDC overlap having just completed its exhaust stroke. Cylinder #1 is at TDC overlap when it has just completed its exhaust stroke and cylinder #6 is at TDC having completed its compression stroke. If the engine is viewed from overhead with the rocker covers removed, you can identify engine cylinder position by observing the valves over a pair of companion cylinders. For the TDC position, the pistons in cylinders #1 and #6 are approaching TDC when the valves over #6 are both closed and they have lash. At this point, you can rock the valves over #1 cylinder back and forth and you cannot rock the valves on the #6 cylinder. Cylinder #1 TDC is also indicated by the fact that the exhaust has just closed and the intake is about to open. This method of orienting engine location is commonly used for valve adjustment.

Answer A is incorrect. #1 and #5 are not companion cylinders.

Answer B is incorrect. #4 and #5 are not companion cylinders

Answer C is incorrect. #3 and #6 are not companion cylinders.

Question #19
Answer A is correct. Only Technician A is correct. A clogged crankcase breather tube can cause high crankcase pressure. Blowby—cylinder gas leakage past the piston rings into the crankcase —is usually associated with low power. To check the crankcase

1. Connect a water manometer (or magnehelic gauge) to the oil dipstick tube.

2. Start and run the engine at idle and observe the reading against OEM specifications.

3. Run the engine at 2,000 rpm and record the reading; compare it to OEM specifications.

Answer B is incorrect. Blowby decreases—not increases—power.

Answer C is incorrect. Only Technician A is correct.

Answer D is incorrect. Only Technician A is correct.

Question #20
Answer A is correct. Only Technician A is correct. Glow plugs are exclusively a cold-start aid for truck diesel engines.

Answer B is incorrect. Currently, very few large diesel engines use glow plugs, as they are best suited to older indirect injected engines. Some medium-duty direct injection diesel engines use glow plugs (GP) that are controlled by the electronic control module (ECM) or the power train control module (PCM). In the ECM/PCM, a glow plug relay control is used to energize the glow plugs for assisting cold engine startup. Engine oil temperature, battery positive voltage (B+), and barometric pressure (BARO) are used by the PCM to calculate glow plug on time and the length of the duty cycle. On time normally varies between 10 and 120 seconds. With colder oil temperatures and lower barometric pressures, the plugs are on longer. If battery voltage is abnormally high, the duty cycle is shortened to extend plug life. The glow plug relay will cycle on and off repeatedly only when there is a system high-voltage condition greater than 16 volts.

Answer C is incorrect. Only Technician A is correct.

Answer D is incorrect. Only Technician A is correct.

Question #21
Answer B is correct. The fuel map is part of the proprietary engine files logged in the ECM and would not be adjusted or rewritten in any customer data programming field.

Answer A is incorrect. Maximum road speed is a customer-set parameter.

Answer C is incorrect. Driver rewards can be set by the customer to reward drivers for fuel-conserving operation.

Answer D is incorrect. Governor type can be programmed in the customer data programming.

Question #22
Answer C is correct. High exhaust back pressure will cause lower engine power, higher exhaust temperatures, and poor combustion, but it should not influence the intake/inlet restriction readings, which are a measure of the restrictive factor of the air cleaner.

Answer A is incorrect. High exhaust back pressure will result in low engine power.

Answer B is incorrect. High exhaust back pressure will cause high exhaust temperature.

Answer D is incorrect. High exhaust back pressure will cause poor combustion.

Question #23
Answer B is correct. Only Technician B is correct. A hold in the windings will occasionally warm during operation to the point where they cannot maintain sufficient magnetism to hold the fuel shut-off in the on position. After a cool down period the hold in the windings cools off and starts to work properly again. The engine needs a shutdown solenoid.

Answer A is incorrect. If the pull in the windings failed, the engine would not die. Also if the pull in the windings failed, the engine would not restart.

Answer C is incorrect. Only Technician B is correct.

Answer D is incorrect. Only Technician B is correct.

Question #24
Answer C is correct. A dial indicator is used.

Answer A is incorrect. An alignment machine is not used to check flywheel housing alignment.

Answer B is incorrect. A micrometer will not check alignment.

Answer D is incorrect. A straightedge and feeler gauge will not check alignment.

Question #25
Answer B is correct. Usually there is no negative lead on a starter relay: check a typical bus cranking circuit schematic. This is the LEAST-Likely item to be removed.

Answer A is incorrect. The positive lead would need to be removed.

Answer C is incorrect. The relay mounting nuts would need to be removed.

Answer D is incorrect. The battery ground cable should be removed first.

Question #26
Answer B is correct. The ground connection should always be last and the boost electrical system should be connected in parallel.

Answer A is incorrect. A series circuit would create a dangerous overvoltage condition.

Answer C is incorrect. The negative cable should be last.

Answer D is incorrect. This step is not necessary.

Question #27
Answer C is correct. The LEAST-Likely part of the procedure is checking a filter pad housing for cracks with magnetic flux equipment. Visual checking is usually all that is required to check a filter pad properly.

Answer A is incorrect. The housing should be visually checked for cracks.

Answer B is incorrect. The gasket surfaces should be checked for nicks.

Answer D is incorrect. The passages should be inspected for obstructions.

Question #28
Answer A is correct. Mechanical unit injectors (MUIs) are actuated by cam profile and an injector train actuated by the cam. A problem in the injector actuation train can cause a misfire on one cylinder. All the other conditions listed as answers will cause fueling problems that will affect more than one engine cylinder.

Answer B is incorrect. A restricted fuel filter would cause low power, but not a single cylinder misfire.

Answer C is incorrect. High fuel pressure would not cause a single cylinder misfire.

Answer D is incorrect. If the buffer screw is out of adjustment, the engine will probably have a low idle surge.

Question #29
Answer C is correct. Both Technicians A and B are correct. Any time an EGR system problem is suspected, the ECM should be checked for the presence of DTCs. Since the ECM has control of the EGR system functions, the ECM can detect most malfunctions and codes may be logged. If the vehicle uses EGR controls that operate using compressed air, sufficient air pressure must be available for proper operation. Usually a minimum of 90 psi must be available to the controls.

Answer A is incorrect. Technician B is also correct.

Answer B is incorrect. Technician A is also correct.

Answer D is incorrect. Both technicians are correct.

Question #30
Answer B is correct. The figure shows a typical vibration damper assembly that bolts to the front of a crankshaft. A vibration damper usually consists of a damper drive or housing and inertia ring. The housing is coupled to the crankshaft. Using springs, rubber, or viscous medium,

it drives the inertia ring at average crankshaft speed. Viscous-type harmonic balancers are most common in truck and bus diesels; the annular housing is hollow and bolted to the crankshaft. Within the hollow housing, the inertia ring is suspended in and driven by silicone gel. The shearing of the viscous fluid film between the drive ring and the inertia ring effect the damping action.

Answer A is incorrect. This is not a flywheel assembly.

Answer C is incorrect. This is not a timing gear assembly

Answer D is incorrect. This is not a water pump.

Question #31

Answer B is correct. Only Technician B is correct. If a lifter base is pitted, a very thorough inspection of the mating lobe must be performed. If the lobe shows damage, the camshaft must be replaced.

Answer A is incorrect. A lifter base is usually either flat or slightly convex; a concave lifter base would be an indication of wear.

Answer C is incorrect. Only Technician B is correct.

Answer D is incorrect. Only Technician B is correct.

Question #32

Answer B is correct. A cylinder head must be removed from the cylinder block to be checked for cracks so checking the torque of the cylinder head bolts would not be part of the procedure as this is only done with the head attached to the block deck.

Answer A is incorrect. Removing gasket material is a step.

Answer C is incorrect. Magnetic crack detection is a good detection method.

Answer D is incorrect. Removing carbon build-up is necessary to find cracks.

Question #33

Answer C is correct. Both Technicians A and B are correct. Injectors may be removed from a cylinder head with an injector heel bar or by either method mentioned by the two technicians.

Answer A is incorrect. Technician B is also correct.

Answer B is incorrect. Technician A is also correct.

Answer D is incorrect. Both technicians are correct.

Question #34

Answer C is correct. Both Technicians A and B are correct. Caterpillar recommends a preliminary test for mechanical unit injectors using an infrared thermometer. When performing this test, the engine should be operated at low idle and the temperature at the exhaust manifold ports measured. Low temperature at one of the ports can indicate no fuel from an injector, while excessively high temperature is evidence of overfueling. The temperature difference between cylinders should not be greater than 70°C (158°F). With the valve cover removed and the engine idling, each unit injector may be tested individually by moving its rack to the fuel-on position. This should immediately result in combustion knock in the cylinder being tested. If this combustion knock does not occur, there may be a problem with the injector assembly, the fuel supply to it, or the seal between the injector and sleeve. Technician A is correct. Checking temperatures isolates the misfiring cylinder on a hydromechanical engine. Technician B is also correct many engine manufacturers have recommended canceling the injector as a means to locate the misfiring cylinder.

Answer A is incorrect. Technician B is also correct.

Answer B is incorrect. Technician A is also correct.

Answer D is incorrect. Both technicians are correct.

Question #35

Answer A is correct. The DMM has been placed in series in the circuit shown, meaning that it has become part of the circuit and full current will pass through it. Amperage or current flow is measured. Ammeters are used to measure the amperage or current that is being pushed through the load to do work. The other type of ammeter used is the inductive pickup type. To measure current flow, the pickup is clamped around the wire in the circuit. This eliminates the need to break into the circuit with the meter. A small winding in the pickup produces a voltage that is proportional to the magnetic field around the conductor when current flows. The voltage measured is calibrated to give an amperage reading. When using a direct-reading ammeter, never place the meter leads across a battery or a load as this can, at a minimum, blow the meter's fuse or possibly destroy the meter.

Answer B is incorrect. Voltage drop is measured with a voltmeter in parallel to the circuit.

Answer C is incorrect. Circuit resistance is measured with an ohmmeter connected to a circuit which is not energized.

Answer D is incorrect. Wattage is measured by measuring both voltage and amperage.

Question #36

Answer D is correct. Neither technician is correct.

Answer A is incorrect. Technician A is incorrect. Cylinder valve cupping means that the valve has failed.

Answer B is incorrect. Technician B is incorrect. The valve margin is one of the most critical indicators of a valve's serviceability. Measure the valve margin—the dimension between the valve seat and the flat face of the valve mushroom. This specification is critical when machining valves, and it must exceed the minimum specified value when the grinding process has been completed. A valve margin that is lower than the specification will result in valve failures caused by overheating.

Answer C is incorrect. Neither technician is correct.

Question #37

Answer C is correct. Both Technicians A and B are correct. Push tubes should never be straightened as they will fail again, usually sooner than later. Checking ball socket integrity is an essential check to perform before returning push tubes to service. Both pushrods and tubes seldom fail under normal operation, but they are vulnerable to inaccurate valve lash adjustments and engine overspeeding. Ball and socket wear can be checked visually. Flaking or disintegration of the hard surface indicates that the tube or rod should be replaced.

Answer A is incorrect. Technician B is also correct.

Answer B is incorrect. Technician A is also correct.

Answer D is incorrect. Both technicians are correct.

Question #38

Answer D is correct. Neither technician is correct. The most likely cause of this problem is a malfunctioning air fuel ratio control (aneroid).

Answer A is incorrect. Technician A is incorrect. A restricted fuel filter will limit fuel delivery. This engine is obviously receiving plenty of fuel if it is making black smoke.

Answer B is incorrect. Technician B is incorrect. A restricted air filter could cause black smoke, however, the engine would be low on power if the air filter was restricted. The question states the engine runs fine otherwise.

Answer C is incorrect. Neither technician is correct.

Question #39

Answer C is correct. Both Technicians A and B are correct. Technician A correctly describes the open to fully open specifications on a typical 9-psi-rated radiator cap. Technician B also correctly describes the differential specification that the radiator cap uses to admit air to the cooling system to prevent radiator hose collapse. Care should be exercised when removing a radiator cap from the radiator. If the system is pressurized, hot coolant may escape from the filler neck with great force. Most filler necks are fitted with double-cap lock stops to prevent the radiator cap from being removed in a single counterclockwise motion. If the radiator is still pressurized, the cap will jam on the intermediate stops. Never attempt to remove a radiator cap until the cooling system pressure is equalized. Radiators are usually equipped with a pressure cap designed to maintain a fixed operating pressure while the engine is running. This cap is also equipped with a vacuum valve to admit surge tank coolant (or air) into the cooling circuit (the upper radiator tank) when the engine is shut down to accommodate coolant thermal contraction. Radiator caps permit pressurization of a sealed cooling system. For each 1 psi (7 kPa) above atmospheric pressure, coolant boil point is raised by 3°F (1.67°C) at sea level. For every 1,000 feet of elevation, the boil point decreases by 1.25°F (0.5°C). System pressures will seldom be designed to exceed 25 psi (172 kPa); more typically they will range between 7 psi (50 kPa) and 15 psi (100 kPa).

Answer A is incorrect. Technician B is also correct.

Answer B is incorrect. Technician A is also correct.

Answer D is incorrect. Both technicians are correct.

Question #40

Answer C is correct. Both Technicians A and B are correct. Piston ring end gaps should be offset on installation to limit blowby. It is important when installing a piston assembly into a cylinder to take measures to protect the studs from damaging the crankshaft journals. Observe OEM instructions. Dividing the number of rings into a 360° circle usually offsets the gaps; so if there were three rings, the stagger would be 120° offset. The ring gaps are usually not placed over the thrust or antithrust side of the piston.

Answer A is incorrect. Technician B is also correct.

Answer B is incorrect. Technician A is also correct.

Answer D is incorrect. Both technicians are correct.

Question #41

Answer B is correct. Only Technician B is correct. Stripped threads can be repaired by several processes and using a helicoil insert is one such method. A common fastening problem is stripped threads, which is usually caused by torque that is too high or by cross-threading. Rather than replacing the block or cylinder head, the threads can be replaced by the use of threaded inserts. Several types of threaded inserts are available—the helically coiled insert is the most popular.

Answer A is incorrect. The block does not need to be replaced.

Answer C is incorrect. Only Technician B is correct.

Answer D is incorrect. Only Technician B is correct.

Question #42

Answer C is correct. Both Technicians A and B are correct. Antifreeze does not evaporate. When it leaks, ram air can spread it over a large area in the engine compartment. A pinhole leak in the radiator or an upper radiator hose could cause the engine to be coated with a film of antifreeze. Cooling system leakage is common, and the system should be inspected by the operator daily. Cold leaks may be caused by contraction of mated components at joints, especially hose clamps. Cold leaks often cease to leak at operating temperatures. Many fleet operators replace all the coolant after a prescribed in-service period regardless of its appearance to avoid the costs incurred in a breakdown. Silicone hoses are more expensive than the rubber compound type but they usually have longer service life. Silicone hoses require the use of special clamps, and these are sensitive to overtightening. They must be torqued to the required specification. Pressure testing a cooling system will locate most external cooling system leaks. A typical cooling system pressure testing kit consists of a hand-actuated pump and gauge assembly plus various adapters for the different types of fill necks and radiator caps. Some are capable of vacuum testing.

Answer A is incorrect. Technician B is also correct.

Answer B is incorrect. Technician A is also correct.

Answer D is incorrect. Both technicians are correct.

Question #43
Answer B is correct. Only Technician B is correct. Bubbles flowing in the fuel water separator indicate suction-side leaks.

Answer A is incorrect. This is not considered normal. Any presence of air in the fuel system can cause poor engine performance, rough running, and can eventually lead to fuel injector and/or fuel pump damage, due to lubrication starvation of internal components.

Answer C is incorrect. Only Technician B is correct.

Answer D is incorrect. Only Technician B is correct.

Question #44
Answer A is correct. A stuck closed oil pressure relief valve would be unable to release and therefore more likely to cause high oil pressure complaints. It would be LEAST-Likely to cause low oil pressure complaints.

Answer B is incorrect. A restricted oil pump suction tube would starve the oil pump of oil, resulting in low oil pressure

Answer C is incorrect. Worn main bearings would reduce resistance to oil flow, causing pressure to drop.

Answer D is incorrect. A clogged filter could cause oil starvation, and low-pressure concerns.

To verify low oil pressure:

- Check the oil sump level.
- Install a master gauge (an accurate, fluid-filled gauge).
- Investigate the oil consumption history.
- Determine the cause.

Some possible causes of low oil pressure conditions are:

- Contaminated lube oil (fuel)
- Excessive crankshaft bearing clearance
- Excessive camshaft or rocker shaft bearing clearance
- Pump relief valve spring fatigued or stuck open
- Oil pump defect
- Oil suction pipe defect
- Aerated oil

Question #45
Answer A is correct. Only Technician A is correct. You may use either an Hg manometer or a pressure gauge.

Answer B is incorrect. To measure boost pressure the engine needs to be run above idle speed and under load to produce boost pressure. This may be performed using a dynamometer, or during a road test.

Answer C is incorrect. Only Technician A is correct.

Answer D is incorrect. Only Technician A is correct.

Question #46

Answer C is correct. In a mechanical unit injector fuel system, each cylinder has a unit injector that receives low-pressure fuel from a gear-type charging pump. A mechanical governor controls the fuel rack, which in turn controls each unit injector. Since this system is mechanical, not electronic, there is no injector driver module.

Answer A is incorrect. MUI does receive fuel from a gear pump.

Answer B is incorrect. MUI is controlled by a mechanical governor.

Answer D is incorrect. Metering is controlled by rack position on most MUI systems.

Question #47

Answer A is correct. Only Technician A is correct. Compressed air is an ideal cleaning medium for the gasket groove in the rocker housing cover.

Answer B is incorrect. Oil may prevent the silicone from properly curing.

Answer C is incorrect. Only Technician A is correct.

Answer D is incorrect. Only Technician A is correct.

Question #48

Answer A is correct. Only Technician A is correct. An improperly installed connecting rod could result in piston skirt cracks. When disassembling an engine, always tag pistons for location on removal, even when the primary cause of the failure has been determined.

Answer B is incorrect. Insufficient piston-to-cylinder liner clearance is more likely to result in skirt scuffing.

Answer C is incorrect. Only Technician A is correct.

Answer D is incorrect. Only Technician A is correct.

Question #49

Answer C is correct. Coolant SCA test strips indicate the relative acidity or alkalinity of the coolant. Acids may form in the coolant when coolant is exposed to combustion gases, when there is corrosion of ferrous and copper metals, and when there is coolant degradation. The coolant may have excess alkalinity when there is aluminum corrosion, and when low-silicate antifreeze is used where high silicate antifreeze is required. Coolant temperature is checked with a thermometer.

Answer A is incorrect. Combustion gases can be identified with test strips.

Answer B is incorrect. Improper antifreeze can be identified with test strips.

Answer D is incorrect. Coolant degradation can be determined with coolant test strips.

Question #50

Answer A is correct. Only Technician A is correct. The shoe should fit perfectly into the base of the liner.

Answer B is incorrect. If the puller shoe is larger than the liner OD, it will grab on the cylinder bore and not permit the liner to be pulled. Cylinder liners (sleeves) should be removed with a puller and adapter plate or shoe.

Answer C is incorrect. Only Technician A is correct.

Answer D is incorrect. Only Technician A is correct.

Question #51

Answer C is correct. Both Technicians A and B are correct. Petroleum jelly will help reduce corrosion on battery cable terminals. Using protective pads and a light coating of grease will also help limit corrosion.

Answer A is incorrect. Technician B Is also correct.

Answer B is incorrect. Technician A is also correct.

Answer D is incorrect. Both technicians are correct.

Question #52

Answer B is correct. The purpose of the exhaust brake is to restrict the exhaust gas flow leaving the engine, creating drag to help slow the vehicle. If the exhaust brake were to be stuck in the on position, engine power would be greatly reduced. If the engine were to be operated in this condition for an extended period of time, the resulting exhaust back pressure would also cause heat to build in the engine eventually overheating it.

Answer A is incorrect. A clogged air filter will certainly reduce engine power; it should not make it overheat.

Answer C is incorrect. Aerated fuel will certainly reduce engine power; it should not make it overheat.

Answer D is incorrect. If the fan were stuck on high speed, it would not reduce engine power and the coolant temperature may actually be lower than normal.

Question #53

Answer C is correct. Both Technicians A and B are correct. One of the functions of a radiator cap is to relieve pressure by discharging fluid to an expansion chamber when it exceeds a specified value, and most diesel cooling systems function at maximum pressures of 15 psi (100 kPa) or less. Radiators are usually equipped with a pressure cap designed to maintain a fixed operating pressure while the engine is running. This cap is additionally equipped with a vacuum valve to admit surge tank coolant (or air) into the cooling circuit (the upper radiator tank) when the engine is shut down to accommodate coolant thermal contraction. Radiator caps permit pressurization of a sealed cooling system. For each 1 psi (7 KPa) above atmospheric pressure, coolant boil point is raised by 3°F (1.67°C) at sea level; for every 1,000 feet of elevation, the boil point decreases by 1.25°F (0.5°C). System pressures are seldom designed to exceed 25 psi (172 KPa), and more typically they range between 7 psi (50 KPa) and 15 psi (100 KPa).

Answer A is incorrect. Technician B is also correct.

Answer B is incorrect. Technician A is also correct.

Answer D is incorrect. Both technicians are correct.

Question #54

Answer C is correct. Both technicians are correct. Using an analog multimeter can damage an electronic system, so a digital multimeter with at least 10-mega ohm impedance is recommended.

Answer A is incorrect. Technician B is also correct.

Answer B is incorrect. Technician A is also correct.

Answer D is incorrect. Both technicians are correct.

Question #55

Answer A is correct. Only Technician A is correct. Turbo boost sensors can shift their values and show a small amount of boost when there is none. The best way to check this is to install a scan tool and check the turbo boost sensor reading with key on engine off. If the ECM suspects a boost under this condition, either the ECM or the sensor/wiring is faulty.

Answer B is incorrect. A throttle position sensor can fail, but if the ECM sees throttle input it will accelerate the engine, not overfuel it.

Answer C is incorrect. Only Technician A is correct.

Answer D is incorrect. Only Technician A is correct.

Question #56

Answer A is correct. Only Technician A is correct. During engine overhaul, all the gears in the engine timing gear train should be thoroughly inspected.

Answer B is incorrect. Rolling or lipping of gear teeth is not acceptable and evidence of either requires the replacement of the gear and its mate(s).

Answer C is incorrect. Only Technician A is correct.

Answer D is incorrect. Only Technician A is correct.

Question #57

Answer B is correct. The most accurate place to obtain a fuel pressure reading is at the delivery port of the fuel pump. The pump is generating the pressure.

Answer A is incorrect. Fuel pressure is not present at the suction filter housing.

Answer C is incorrect. There is no fuel pressure measurable in the tank.

Answer D is incorrect. Fuel pressure is not present at the pump inlet.

Question #58

Answer D is correct. Ground wires should be attached to an unpainted section of the frame. All of the other answers are conditions that could cause electrical integrity problems.

Answer A is incorrect. Loose or improperly mated connectors can cause a problem in an electrical/electronic circuit.

Answer B is incorrect. Wires with faulty insulation can cause a problem in an electrical/ electronic circuit.

Answer C is incorrect. A melted or distorted electrical connection can be an indication of a high-resistance connection, and a likely cause of the problem.

Question #59

Answer B is correct. A defective injector will cause fuel system problems, but it is unlikely this would affect the cooling system so this would be the LEAST-Likely cause of the surging condition.

Answer A is incorrect. A loose belt could cause coolant-level surge.

Answer C is incorrect. An internally restricted radiator could cause coolant-level surging.

Answer D is incorrect. A blown head gasket can let compression gases in the cooling system, which would result in surging coolant.

Question #60

Answer B is correct. Battery capacity means battery potential, which is potential difference or voltage. Voltage is measured with a voltmeter. The location of batteries in some buses makes it physically impossible to use a hydrometer to check the state of charge unless the batteries are pulled. Also, some batteries are totally sealed. An open circuit voltage test is especially useful in these circumstances.

Answer A is incorrect. A voltmeter is the best way to check battery potential.

Answer C is incorrect. An ohmmeter cannot measure battery potential.

Answer D is incorrect. An ammeter cannot measure battery potential.

Question #61

Answer A is correct. Only Technician A is correct. Flywheel ring gears are usually heat shrunk to flywheels.

Answer B is incorrect. Technician B's statement that a flywheel should be replaced when the ring gear teeth are damaged is incorrect. Shrunkfit to the outer periphery of the flywheel is the ring gear, which provides the means of transmitting cranking torque to the engine by the starter motor during startup. To service the ring gear, first remove the flywheel from the engine. Using an oxyacetylene torch, partially cut through the ring gear working from the outside and remove the ring gear from the flywheel. In most cases this is sufficient to expand the ring gear so that it can be removed using a hammer and chisel. Avoid heating the flywheel any more than absolutely necessary or damaging the flywheel by careless use of the oxyacetylene flame.

Answer C is incorrect. Only Technician A is correct.

Answer D is incorrect. Only Technician A is correct.

Question #62
Answer B is correct. Only Technician B is correct. Voltage drop testing identifies high resistances in energized circuits.

Answer A is incorrect. Resistance cannot be checked in an energized circuit and would damage the ohmmeter using this method. An ohmmeter provides a means of effectively isolating parts of electrical circuits and components to pinpoint problems. The meter can accomplish this because it provides its own power supply to the load or circuit being tested and provides a return back to its internal power source. The meter becomes part of the circuit and provides the power.

Answer C is incorrect. Only Technician B is correct.

Answer D is incorrect. Only Technician B is correct.

Question #63
Answer C is correct. Elevated oil levels in an engine may cause the rotating internal parts to churn and force their way through the oil. This can generate excessive heat and rob power from the engine, causing performance complaints. It also will introduce air bubbles into the oil. Aerated oil can cause severe engine damage due to oil starvation of components. Overfilling the engine oil will not cause high oil pressure; oil pressure is produced by the oil pump and controlled by the relief/regulator valves.

Answer A is incorrect. Engine overheating can occur.

Answer B is incorrect. Aerated oil can occur.

Answer D is incorrect. Low power complaints can occur.

Question #64
Answer B is correct. A water pump weep hole may produce occasional drops of coolant when the pump is functioning properly. All the other conditions listed can cause cooling system failures because they would cause loss of circulation, loss of coolant, or inhibit the cooling capability of the engine.

Answer A is incorrect. The described belt should be changed.

Answer C is incorrect. A loose hose clamp should be fixed.

Answer D is incorrect. Rust or scum in the cooling system can cause overheating, and may be an indication of more serious engine problems.

Question #65
Answer A is correct. Only Technician A is correct. Cylinder block warpage is correctly measured with a straightedge and thickness gauges.

Answer B is incorrect. A dial indicator is not used to check block deck warpage. Technician B's method of measuring a block deck with a dial indicator is incorrect. A typical maximum specification will be in the range of 0.004–0.005 in. (0.10–0.127 mm) overall, and about 0.003 in. (0.076 mm) side to side. Readings beyond this specification indicate the need to resurface the block fire deck. However, you should always refer to the OEM service manual for the correct value. Blocks do not warp nearly as often as cylinder heads. When the block fire deck is warped or not parallel to the main bearing bores, it may be resurfaced with a milling machine or grinder. Next, check the main bearing bore and alignment. It is good shop practice to record the amount of stock removed from the deck by stamping the amount removed on the cylinder block pad.

Answer C is incorrect. Only Technician A is correct.

Answer D is incorrect. Only Technician A is correct.

Question #66
Answer A is correct. If the engine speed sensor fails, the computer can loose its timing reference; in effect, the ECM no longer knows where the crankshaft is so it cannot operate the injectors. Sometimes this failure will not set a diagnostic trouble code (DTC). The ECM does not know the engine is rotating so it does not know the sensor has failed.

Answer B is incorrect. A stuck open wastegate would not cause this condition.

Answer C is incorrect. A clogged air filter would not cause this condition.

Answer D is incorrect. A failed temperature sensor would not cause this condition.

Question #67
Answer A is correct. The fuel system shown is a rotary distributor pump fueling an 8-cylinder engine. This is determined by the 8 high-pressure lines that route the fuel from the hydraulic head to the injectors. The distributor-type fuel injection pump pressurizes and distributes a metered amount of fuel to each cylinder nozzle at the proper time based on the calibrated needs of the engine. The distributor-type uses one pump barrel and a set of plungers to supply all cylinders in rotation.

Answer B is incorrect. This is not a HEUI fuel system. HEUI uses individual injectors.

Answer C is incorrect. This is not an in-line pump. The use of a separate pumping element for each cylinder has been the general practice of the in-line pump.

Answer D is incorrect. This is not a unit injector fuel system.

Question #68
Answer A is correct. Only Technician A is correct. A leaking thermostat seal could cause coolant to bypass a closed thermostat, resulting in an engine that would fail to reach operating temperature.

Answer B is incorrect. A loose fan belt would most likely cause the engine to overheat.

Answer C is incorrect. Only Technician A is correct.

Answer D is incorrect. Only Technician A is correct.

Question #69
Answer A is correct. Weatherproof electronic connectors must be serviced using the proper insertion and removal tooling; the other answers outline practices that are incorrect.

Answer B is incorrect. The pins must not be bent.

Answer C is incorrect. Weatherproof connectors should only be replaced with weatherproof connectors.

Answer D is incorrect. Wiring does not need to be grounded prior to changing.

Question #70
Answer B is correct. Only Technician B is correct. Most hydromechanical engines will produce some visible exhaust smoke during cold startup while electronically controlled engines are programmed to minimize this until temperatures drop to below 0°F (−17°C). Technician B correctly identifies the startup smoke condition and is also correct in suggesting that a block immersion heater would help cold starts.

Answer A is incorrect. A broken injector nozzle spring would usually cause smoking and misfire throughout the operating range.

Answer C is incorrect. Only Technician B is correct.

Answer D is incorrect. Only Technician B is correct. When a diesel engine is running properly with a heavy load, its exhaust is almost clear. Oil adds a blue haze to the exhaust, causing a blue smoke. Unburned fuel or an improper grade of fuel produces a gray or black smoke; water or coolant will produce white smoke-like steam. White smoke may also indicate that injectors are misfiring. White smoke occurs either when the engine first starts or when the ambient temperature is too low for proper combustion. White smoke occurs more often in an indirect injected (IDI) engine from retarded timing. Very late timing on a direct injected (DI) engine also causes white smoke. Gray or black smoke is the result of incomplete combustion. Air starvation is the major cause of black smoke.

Question #71
Answer A is correct. All of the conditions listed in the answers have the potential to cause a coolant in oil condition (depending on the specific engine), EXCEPT a crack in an exhaust valve seat. Exhaust valve seats are usually separate from cylinder head castings, so a crack would affect the exhaust seat only. Coolant in the engine lubricant gives it a milky, cloudy appearance when churned into the oil. After settling, the coolant usually collects at the bottom of the oil sump. When the drain plug is removed, the heavier coolant exits first as long as sufficient time has passed since running the engine. When coolant is found in the engine oil, the cylinder head(s) and cylinder head gasket are the most likely source.

Answer B is incorrect. A crack in the water jacket would cause oil contamination.

Answer C is incorrect. An internal leak on a water pump could cause oil contamination if the pump is gear driven.

Answer D is incorrect. A leaking head gasket can cause oil contamination.

Question #72
Answer B is correct. The diagram shows a typical chassis-mounted charge-air cooler that uses ram air as the cooling medium. Ram air is the airflow created as a vehicle moves along the roadway. Air-to-air coolers have the appearance of a coolant radiator and are often chassis-mounted in front of the radiator. As the vehicle moves down the roadway, ambient air is forced through the fins and element tubing. Ram air is therefore the cooling medium. Cooling efficiencies are highest when the vehicle is traveling at higher speeds.

Answer A is incorrect. An air-to-air aftercooler is shown.

Answer C is incorrect. This is not a tip turbine-type air cooler.

Answer D is incorrect. Exhaust gas routed through the charge-air cooler would warm the air not cool it.

Question #73
Answer B is correct. Only Technician B is correct. Technician B's idea of talking to the operator to get more information about the excessive smoke condition is a good one.

Answer A is incorrect. There are many causes of excessive smoking in a diesel engine and many of the conditions can be repaired without overhauling the engine.

Answer C is incorrect. Only Technician B is correct.

Answer D is incorrect. Only Technician B is correct.

Question #74
Answer C is correct. The LEAST-Likely tool to be used to detect this condition would be a feeler gauge.

Answer A is incorrect. An inside micrometer can be used to detect a cylinder out-of-round condition.

Answer B is incorrect. A telescoping gauge can be used to detect a cylinder out-of-round condition.

Answer D is incorrect. A dial bore gauge can be used to detect a cylinder out-of-round condition.

Question #75
Answer D is correct. Neither technician is correct. Back-flushing of cooling systems is unnecessary if the coolant is maintained and conditioned according to the manufacturer's instructions. It is correct procedure to refill the cooling system with any bleed valves in the system open as this allows air to purge as the coolant enters the system. Most commercially available in-chassis radiator descaling solutions are a poor risk, because they are seldom 100 percent effective and may dislodge scale, which will subsequently plug up elsewhere in the cooling circuit. For the same reasons, reverse-flow flushing of the cooling system makes little sense. When OEMs recommend radiator flushing, it is generally performed in the normal direction of flow, often aided by a cleaning solution.

Answer A is incorrect. Technician A is incorrect.

Answer B is incorrect. Technician B is incorrect.

Answer C is incorrect. Neither technician is correct.

Question #76
Answer A is correct. While a high crankcase oil level will cause other problems, it is LEAST-Likely to result in cracks and scoring of the piston skirts. The other three problems would be capable of damaging the piston skirts.

Answer B is incorrect. Improper piston clearance can cause damaged piston skirts.

Answer C is incorrect. Overfueling can cause piston damage.

Answer D is incorrect. Engine overheating can cause piston damage.

Question #77
Answer A is correct. Service bulletins provide the latest service tips, field repairs, product improvements, and related information of benefit to service personnel. Some service bulletins are updates to information in the service manual; these take precedence over service manual information, until the latter is updated. At that time, the bulletin is usually canceled. The service bulletins manual is usually available only to dealers who in turn supply the owners of the vehicle. When doing service work on a vehicle system or part, check for a valid service bulletin for the latest information on the subject. Because technical service bulletins are more frequently released, they take precedence over previously published information on the same subject.

Answer B is incorrect. Service manuals are subject to being out of date due to the time required from writing to printing to distribution.

Answer C is incorrect. Parts manuals do not contain service information.

Answer D is incorrect. Typically, operators manuals only list owner service information.

Question #78
Answer C is correct. Both Technicians A and B are correct. When the liner checks within serviceability specifications, the technician deglazes the liner. Deglazing removes the least amount of material using a power-driven flex hone or rigid hone with 200–250-grit stones. The best type of glaze buster is the flex hone, typically a conical (Xmas tree) or cylindrically shaped shaft with flexible branches with carbon/abrasive balls. The objective of glaze busting is to machine away the cylinder ridge above the ring belt travel and reestablish the crosshatch. Set the drill at 120–180 rpm and use rhythmic, reciprocating thrusts at short sequences; stopping frequently to inspect the finish produces the best results. You should observe a 60° crossover angle, 20–35 microinch crosshatch. Technician A is correct in describing the procedure required to produce a crosshatch in the liner. Technician B is also correct in saying that soap and water is the preferred method of cleaning a liner after honing.

Answer A is incorrect. Technician B is also correct.

Answer B is incorrect. Technician A is also correct.

Answer D is incorrect. Both technicians are correct.

Question #79

Answer B is correct. Only Technician B is correct. High-pressure lines can be removed individually when required, providing the clamps are first removed.

Answer A is incorrect. The injectors lines do not have to be removed as a unit.

Answer C is incorrect. Only Technician B is correct.

Answer D is incorrect. Only Technician B is correct.

Question #80

Answer C is correct. All of the conditions in the answers would be capable of causing low power, but in approaching this problem in a hydromechanical engine, the first step would be to check the accelerator linkage.

Answer A is incorrect. A restriction in the crossover line would most likely cause unequal fuel levels between the tanks.

Answer B is incorrect. The throttle arm break over setting would not be the first check.

Answer D is incorrect. Although governor settings can cause low power, they would not be the first thing to check.

Question #81

Answer D is correct. If the throttle position sensor was faulty, the ECM would not recognize that the operator had increased throttle position from idle and therefore would not "know" to turn off the engine brake.

Answer A is incorrect. If the clutch micro switch was faulty, the engine brake would Most-Likely not disengage when changing gears.

Answer B is incorrect. If the master switch was faulty, the engine brake would Most-Likely not engage at all.

Answer C is incorrect. If the service brake switch was faulty, the engine brake would not work in conjunction with the service brake when programmed to do so in the ECM.

Question #82

Answer A is correct. Only Technician A is correct. A leak in an air hose upstream from the turbocharger compressor can allow unfiltered air to be pulled in. Abrasives in the air can wear the compressor wheel vanes to knife-edge sharpness.

Answer B is incorrect. The aftercooler is located downstream from the compressor housing.

Answer C is incorrect. Only Technician A is correct.

Answer D is incorrect. Only Technician A is incorrect.

Question #83

Answer B is correct. Only Technician B is correct. Excessive manifold boost pressures are the result of high turbine speeds. A stuck turbocharger wastegate could cause this condition by not diverting exhaust gas flow around the turbine.

Answer A is incorrect. A punctured air filter would allow some unfiltered air to enter the engine, but would not cause high manifold boost.

Answer C is incorrect. Only Technician B is correct.

Answer D is incorrect. Only Technician B is correct.

Question #84

Answer A is correct. Operation of a diesel engine under full load for prolonged periods is exactly what a diesel engine is designed for and will not create problems as long as the engine is prevented from overheating. A small percentage of crankshaft failures will result from manufacturing and design problems. Design problems usually only occur shortly after an engine series is introduced. No OEM wants to have an engine labeled with a basic engineering fault, so design problems are usually rectified quickly.

Answer B is incorrect. An unbalanced vibration damper can cause crankshaft failure.

Answer C is incorrect. An incorrectly installed flywheel housing can cause crankshaft and bearing damage.

Answer D is incorrect. Loose torque converter bolts can damage the crankshaft.

Question #85

Answer C is correct. Both technicians are correct. Main bearings tell a story, so they should be checked on disassembly as should the crankshaft for obvious signs of failure. After cleaning, the crankshaft should again be subject to measuring to determine its serviceability. Lubrication-related failures are caused by insufficient or complete absence of oil in one or all the crank journals. The bearing is subjected to high friction loads and surface welds itself to the affected crank journal. This may result in a spun bearing in which the bearing rotates with the journal in the bore or continues to scuff the journal until it is destroyed. When a crankshaft fractures as a result of bearing seizure, the surface of the journal is destroyed by excessive heat; it fails because it is unable to sustain the torsional loading.

Answer A is incorrect. Technician B is also correct.

Answer B is incorrect. Technician A is also correct.

Answer D is incorrect. Both technicians are correct.

Question #86

Answer C is correct. Both Technicians A and B are correct. Turbochargers are lubricated by engine oil, so Technician A is correct. Technician B is also correct because when turbine bearings fail, fragments can be distributed into the engine oil sump, so the oil should be changed.

Answer A is incorrect. Technician B is also correct.

Answer B is incorrect. Technician A is also correct.

Answer D is incorrect. Both technicians are correct.

Question #87

Answer D is correct. Plunger-type fuel transfer pumps can be equipped with a sediment bowl, a sediment strainer, and a hand-primer pump, but do not come with pop-off valves. The charging pressure produced by these pumps is usually defined downstream from the pump itself.

Answer A is incorrect. They can have a sediment bowl.

Answer B is incorrect. They can have a sediment strainer.

Answer C is incorrect. They can have a hand primer.

Question #88

Answer C is correct. Both Technicians A and B are correct. Inductive-pulse generators are used to signal road speed values to the electronic systems. The road speed sensor is usually located in the tail shaft of the transmission.

Answer A is incorrect. Technician B is also correct.

Answer B is incorrect. Technician A is also correct.

Answer D is incorrect. Both technicians are correct.

Question #89
Answer A is correct. Only Technician A is correct. Pairs of belts should always be replaced as a set. Belts should be adjusted using a belt tensioner. The consequences of maladjusted belts are:

• Too tight: Excessive loading of the bearings shortens bearing and belt life.

• Too loose: Slippage destroys belts even more rapidly than a too tight adjustment.

Belts should be inspected periodically as part of a PM routine. Replace belts if they are glazed, cracked, or nicked. Replacing belts with early indications of failure costs much less in the long run than the breakdowns that may occur when belts fail in service.

Answer B is incorrect. This method of installing belts is not recommended as damage to belts and even to the pulleys can result.

Answer C is incorrect. Only Technician A is correct.

Answer D is incorrect. Only Technician A is correct.

Question #90
Answer A is correct. Only Technician A is correct. Worn bushings increase clearances which reduce oil pressure.

Answer B is incorrect. When camshaft bushings are removed, they cannot be reinstalled. Once they are driven out they will not stay snug fit when reinstalled.

Answer C is incorrect. Only Technician A is correct.

Answer D is incorrect. Only Technician A is correct.

Question #91
Answer D is correct. To perform a voltage drop test, the voltmeter test leads are placed in parallel to the circuit being tested.

Answer A is incorrect. Voltage drop is not checked across the battery terminals. Battery voltage is checked across the battery terminals.

Answer B is incorrect. This check would be measuring battery voltage.

Answer C is incorrect. Amperage is measured when the leads are put in series with the circuit.

Question #92
Answer B is correct. Only Technician B is correct. Piston ring end gap should be measured by placing the ring into the cylinder alone and by measuring the gap with thickness gauges. This is always done before fitting the ring to the piston. Check the ring gap by installing a new ring into the cylinder bore, and measure it with a thickness (feeler) gauge.

Answer A is incorrect. It would be impossible to measure the ring end gap after the assembly is in the engine.

Answer C is incorrect. Only Technician B is correct.

Answer D is incorrect. Only Technician B is correct.

Question #93
Answer D is correct. The procedure involves the pressure testing of the fuel delivery system in a cylinder head so test injectors, compressed air, and threaded plugs to seal off the galleries are all used. The injection pump would not play a role in this test procedure so this would be the item LEAST-Likely to be used.

Answer A is incorrect. Test injectors are used to fill the injector bore and hold the injector sleeves in place.

Answer B is incorrect. Compressed air is used to check the passages.

Answer C is incorrect. Threaded plugs are used to block passages.

Question #94

Answer C is correct. A HEUI will *not* use a hydro mechanical governor. The HEUI is a hydraulically actuated electronic unit injector fuel system.

Answer A is incorrect. A HEUI does use a transfer pump.

Answer B is incorrect. A HEUI does use unit injectors.

Answer D is incorrect. A HEUI does use a plunger-style oil pump.

Question #95

Answer B is correct. Only Technician B is correct. Improperly installed keepers could be the cause of dropping a valve after an engine overhaul.

Answer A is incorrect. Improperly installed valve seals can cause excessive oil consumption but are unlikely to be the cause of dropping a valve after an engine overhaul.

Answer C is incorrect. Only Technician B is correct.

Answer D is incorrect. Only Technician B is correct.

Question #96

Answer C is correct. Both Technicians A and B are correct. Bent, cracked, or otherwise damaged fan blades can cause engine imbalance and the cost of repair can be considerably greater than the cost of a fan assembly. Technician A is of the opinion that a damaged fan should be replaced and Technician B says that a bent fan blade can cause engine vibration. Both are correct.

Answer A is incorrect. Technician B is also correct.

Answer B is incorrect. Technician A is also correct.

Answer D is incorrect. Both technicians are correct.

Question #97

Answer A is correct. SAE MID or message identifiers are used to identify each major electronic circuit on a vehicle chassis. The engine MID is 128 and the transmission MID is 130. SIDs, PIDs, and FMIs all flow from the MID. MID (message identifier) is used to describe a major vehicle electronic system, usually with independent processing capability.

Answer C is incorrect. PID (parameter identifier) is used to code components within an electronic subsystem.

Answer B is incorrect. SID (subsystem identifier) is used to identify the major subsystems of an electronic circuit. SIDs listed here relate to cruise control.

Answer D is incorrect. FMI (failure mode identifier) lists the type of failure (i.e., open, short, out of range, etc.).

Question #98

Answer B is correct. The temperature in the question is 50°F. All battery test specifications are temperature dependent, so it is important to interpret this type of chart. This test is referred to as a load test. A load is placed on the battery to see how it functions under a demand condition. Prior to performing this test, the previous hydrometer and/or open circuit voltage test needs to be performed. If the state of charge is found to be low, the battery has to be charged for the load test to be accurate. A battery tester with an adjustable carbon pile is used to perform this test. The tester usually has an ammeter to indicate the load placed on the battery and a voltmeter to indicate the result in volts. The result and specific numbers required for the test can be summed up as: A good 12-volt battery should maintain a voltage level of 9.6 volts or more when a load of one-half of its cold cranking amps (CCA) is applied to it for 15 seconds.

Answer A is incorrect. Minimum voltage is 9.4 volts at 50°F.

Answer C is incorrect. Minimum voltage is 9.4 volts at 50°F.

Answer D is incorrect. Minimum voltage is 9.4 volts at 50°F.

Question #99

Answer C is correct. Because the question tells you that the engine is operating well and the gauge reading is 0, the cause of the problem is likely an electrical open condition in the gauge/sensor circuit. The other answers describe conditions that are unlikely to result in a 0 reading although they would produce inaccurate/erratic readings. A piezoresistive sensor is the type of sender often used for oil pressure gauges. An ohmmeter is used to check this type of sender by connecting the leads to the sending unit terminal and ground. First, check the resistance with the engine off and compare to specifications. Next, start the engine and allow it to idle. Recheck the resistance value and compare it to specifications. Do not condemn this electrical sending unit immediately if the readings are not within specifications. First, connect a mechanical pressure gauge to the engine to confirm that it is producing adequate oil pressure.

Answer A is incorrect. The oil passage should be open.

Answer B is incorrect. The wrong oil Most-Likely would not cause the gauge to read 0. It may read low but not 0.

Answer D is incorrect. If the engine was not at operating temperature, the oil pressure would most likely be higher than normal, causing a high gauge reading, not 0 pressure.

Answers to the Test Questions for the Additional Test Questions Section 6

1. C	35. C	69. C	103. C
2. D	36. A	70. D	104. A
3. B	37. C	71. C	105. D
4. A	38. A	72. B	106. C
5. D	39. B	73. D	107. C
6. C	40. B	74. A	108. C
7. A	41. A	75. A	109. A
8. A	42. C	76. D	110. A
9. A	43. C	77. C	111. A
10. C	44. C	78. D	112. C
11. C	45. C	79. A	113. C
12. D	46. A	80. D	114. B
13. A	47. D	81. B	115. B
14. C	48. A	82. D	116. C
15. C	49. B	83. C	117. A
16. D	50. A	84. A	118. A
17. D	51. D	85. D	119. A
18. B	52. D	86. C	120. B
19. B	53. C	87. B	121. C
20. D	54. D	88. D	122. A
21. C	55. A	89. A	123. A
22. C	56. B	90. C	124. A
23. B	57. B	91. B	125. D
24. A	58. A	92. C	126. A
25. D	59. C	93. B	127. C
26. D	60. D	94. B	128. A
27. C	61. C	95. C	129. D
28. A	62. C	96. A	130. C
29. D	63. C	97. B	131. A
30. A	64. B	98. B	132. A
31. C	65. B	99. B	133. D
32. B	66. D	100. B	134. C
33. A	67. A	101. C	135. B
34. B	68. B	102. B	136. C

137. B	153. C	169. C	185. C
138. D	154. D	170. D	186. B
139. C	155. B	171. A	187. C
140. D	156. A	172. D	188. C
141. C	157. B	173. C	189. D
142. A	158. B	174. D	190. B
143. C	159. C	175. B	191. B
144. C	160. B	176. C	192. A
145. D	161. A	177. B	193. C
146. B	162. D	178. A	194. C
147. A	163. A	179. B	195. B
148. C	164. C	180. C	196. B
149. A	165. A	181. D	197. D
150. A	166. C	182. D	198. B
151. C	167. C	183. A	
152. C	168. C	184. B	

Explanations to the Answers for the Additional Test Questions Section 6

Question #1

Answer C is correct. Both Technicians A and B are correct. Any component after replacement should be tested. Whenever a distributor-type fuel pump is opened up, the gasket surfaces should be cleaned and the components reassembled with new gaskets.

Answer A is incorrect. Technician B is also correct.

Answer B is incorrect. Technician A is also correct.

Answer D is incorrect. Both technicians are correct.

Question #2

Answer D is correct. Voltage dropped overcoming resistance at the solenoid switch contact should be as low as possible, about 0.3 volts, leaving the majority of available voltage for the starter motor.

Answer A is incorrect. This would be excessive voltage drop.

Answer B is incorrect. This would be excessive voltage drop.

Answer C is incorrect. The choice of 0.01 volts is so low that it is unlikely to even be possible.

Question #3

Answer B is correct. Friction-bearing shells must be able to dissipate the heat they are subjected to in the bore into which they are inserted. Bearing crush helps ensure good heat transfer by making it possible to transfer heat. When bearing crush is insufficient, there is less force loading the bearing into its bore and it does not transfer heat as efficiently. Dirt or grease on the back of the bearing would have the same effect. Bearings are retained primarily by crush. The outside diameter of a pair of uninstalled bearing shells slightly exceeds the bore to which it is installed. This creates radial pressure that acts against the bearing halves and provides good heat transfer. The bearing halves are also slightly elliptical to allow the bearing to be held in place during installation; this is known as bearing spread. Tangs in bearings are inserted into notches in the bearing bore to minimize longitudinal movement, prevent bearing rotation, and align oil holes.

Answer A is incorrect. Excessive play would not cause the back of the bearing to be discolored.

Answer C is incorrect. A belled crankshaft journal would establish a wear pattern on the bearing front, not back.

Answer D is incorrect. Degraded lube would establish wear on the front of the bearing, not the back.

Question #4

Answer A is correct. This is the LEAST-Likely symptom of a governor problem.

Answer B is incorrect. Governor response timing can cause high idle overrun.

Answer C is incorrect. Low idle underrun is normally a symptom of a governor problem.

Answer D is incorrect. Hunting can be caused by governor differential lever slop.

Question #5

Answer D is correct. The camshaft main journal runout is being indicated while the shaft is rotated on V-blocks. Install the camshaft in V-blocks and check for shaft bending using a dial indicator.

Answer A is incorrect. The intake lobe is not being measured.

Answer B is incorrect. The exhaust lobe is not being measured.

Answer C is incorrect. Camshaft end play is measured with the cam installed.

Question #6

Answer C is correct. Both Technicians A and B are correct. A leak in the exhaust manifold can affect performance in a turbocharged diesel engine so Technician A is correct. Leaks upstream from the turbo compressor housing can damage the impellor by allowing abrasives to enter the air intake so Technician B is also right.

Answer A is incorrect. Technician B is also correct.

Answer B is incorrect. Technician A is also correct.

Answer D is incorrect. Both technicians are correct.

Question #7

Answer A is correct. Only Technician A is correct. O-rings on wet liners should be changed any time the sleeve is pulled and reinstalled.

Answer B is incorrect. The manufacturer's recommendation for coating wet sleeve O-rings should be observed. This could mean nothing at all, soap, antifreeze, and usually not anaerobic silicone.

Answer C is incorrect. Only Technician A is correct.

Answer D is incorrect. Only Technician A is correct.

Question #8

Answer A is correct. Only Technician A is correct. High crankcase pressures can be caused by a plugged oil breather.

Answer B is incorrect. Technician B is incorrect in saying that all diesel engines use exhaust gas recirculation (EGR), although some do and more will as nitrides of oxygen (NO_x) emissions continue to be a problem for truck diesel engine designers.

Answer C is incorrect. Only Technician A is correct.

Answer D is incorrect. Only Technician A is correct.

Question #9

Answer A is correct. Only Technician A is correct. Voltage drop testing is a means of testing the condition of wires, cables, connections, and loads in the electrical circuit. It is a much better method of testing electrical circuit resistances because the testing is performed in a fully energized electrical circuit as opposed to checks with an ohmmeter, which must be performed on inactive circuits.

Answer B is incorrect. An ammeter can only produce a reading when connected in series in an energized circuit.

Answer C is incorrect. Only Technician A is correct.

Answer D is incorrect. Only Technician A is correct.

Question #10

Answer C is correct. Both Technicians A and B are correct. Calibration codes (bench fuel rates) are indicated on replacement EUIs. These codes should be programmed to the engine management ECM each time they are replaced. EUIs are graded with a calibration code, which enables the ECM to be programmed with data concerning how a specific EUI flows fuel. The diagnostic data reader (DDR) is used to program this data to the ECM and enables highly accurate and balanced fueling of the engine. To program the calibration code, the injector cylinder number would be selected first followed by entering the digital calibration code value.

Answer A is incorrect. Technician B is also correct.

Answer B is incorrect. Technician A is also correct.

Answer D is incorrect. Both technicians are correct.

Question #11
Answer C is correct. Both Technicians A and B are correct.The relative movement between the intake system components on the chassis and those on the engine must be accommodated by having flexible rubber couplings. The smallest leak in the intake system components can allow dirt ingestion by the turbocharger and engine so the turbocharger should first be inspected. If the turbocharger is found to be damaged, the engine should also be inspected.

Answer A is incorrect. Technician B is also correct.

Answer B is incorrect. Technician A is also correct.

Answer D is incorrect. Both technicians are correct.

Question #12
Answer D is correct. If a radiator cap is tested to be defective, replace the cap, test the coolant, and check the radiator hoses for leakage and collapsing. Attempting to adjust or replace the spring is not good practice.

Answer A is incorrect. The cap should be replaced.

Answer B is incorrect. The coolant should be tested.

Answer C is incorrect. The radiator hoses should be checked.

Question #13
Answer A is correct. Only Technician A is correct. Loss of engine oil through a breather tube is a common indication of high crankcase pressure.

Answer B is incorrect. Most current engines run with a specified crankcase pressure and, providing it is within specification, no problems are indicated. Because rings create the seal of a piston in a cylinder bore, some blowby is inevitable, due to end-gap requirement. Normally, blowby is a specification determined by the OEM. Piston rings seal most efficiently when cylinder pressures are highest, and engine rpm is high when there is less time for gas blowby.

Some causes of high crankcase pressure are:

• Cracked head or piston

• Worn, stuck, or broken piston rings

• Glazed liner/sleeve inside wall

• Air compressor (plugged discharge line, air pumped through oil return line)

• Turbocharger (seal failure allows turbine housing to leak pressurized air through to the turbine shaft and back through the oil return piping

Answer C is incorrect. Only Technician A is correct.

Answer D is incorrect. Only Technician A is correct.

Question #14
Answer C is correct. Both Technicians A and B are correct. Good failure analysis is key to avoiding repeat failures. Technician A's method of organizing disassembled pistons is good practice. Improper injectors can cause a variety of piston failures, so Technician B is also correct.

Answer A is incorrect. Technician B is also correct.

Answer B is incorrect. Technician A is also correct.

Answer D is incorrect. Both technicians are correct.

Question #15
Answer C is correct. Both Technicians A and B are correct. The valve bridge must be adjusted before the valve clearance is set, since adjusting the bridge could affect the valve clearance. When torquing the bridge locknuts, they should be removed from the engine and secured in a vice to prevent damage to the bridge guide pins and valves.

Answer A is incorrect. Technician B is also correct.

Answer B is incorrect. Technician A is also correct.

Answer D is incorrect. Both technicians are correct.

Question #16
Answer D is correct. Neither Technician is correct. If the pressure relief valve were stuck closed, pressure would build, possibly to dangerous levels. When pressure testing a cooling system, it makes no difference whether the engine is running or not.

Answer A is incorrect. Technician A is incorrect. The relief valve would have to be stuck open for the system to not maintain pressure.

Answer B is incorrect. Technician B is incorrect. The engine does not have to be running to pressurize the cooling system.

Answer C is incorrect. Neither Technician is correct.

Question #17
Answer D is correct. Neither technician is correct.

Answer A is incorrect. A crack in the counterbore flange of a wet sleeve is reason to replace it, so Technician A is incorrect.

Answer B is incorrect. Technician B is incorrect. It is possible the wet sleeve flange has cracked and produced no damage to the block counterbore. The block counterbore should be carefully measured and inspected.

Answer C is incorrect. Neither technician is correct.

Question #18
Answer B is correct. Only Technician B is correct. All current electronically managed engines have programmed cold-start strategies that limit engine smoking during start and warmup.

Answer A is incorrect. An engine that produces white smoke for 10 minutes after startup is not functioning properly.

Answer C is incorrect. Only Technician B is correct.

Answer D is incorrect. Only Technician B is correct.

Question #19
Answer B is correct. When determining if camshaft bearings are suitable for reuse, the inside diameter should be checked against the manufacturer's specifications for wear. The correct tool for this measurement is an inside micrometer.

Answer A is incorrect. A feeler gauge is used to measure clearance between parts.

Answer C is incorrect. Plastigage is used to measure clearance between parts.

Answer D is incorrect. A manometer is used to measure pressure.

Question #20
Answer D is correct. Neither technician is correct. Vehicle engine and chassis manufacturers adhere to SAE J1587/1708 and J1939 hardware and communications protocols to provide common access to electronic readouts and diagnostic codes by any instrumentation that is SAE compatible. All vehicle manufacturers use either a 6- or 9-pin ATA connector. Multiplexing standardization for vehicles has been orchestrated by ISO (International Organization for Standardization) internationally and by the SAE (Society of Automotive Engineers) in North America. In North America, SAE J standards, J1850 (light-duty vehicles), and J1939 (heavy-duty highway vehicles) define the hardware and software protocols of multiplexed components and data transfer. ISO CAN 2.0 is consistent with J1850 and J1939 standards used in the United States and Canada. The SAE J1939 combines and replaces the older J1587/1708 hardware and software protocols. Current vehicles tend to be both J1939 and J1587/1708 compatible.

Answer A is incorrect. Technician A is incorrect.

Answer B is incorrect. Technician B is incorrect.

Answer D is incorrect. Neither technician is correct.

Question #21

Answer C is correct. To perform an alternator amperage output test, an amp clamp is placed around the wire attached to the alternator's output terminal, engine rpm is increased to high idle, and a load is gradually placed across the battery terminals with a carbon pile load tester until maximum amperage output is reached without reducing battery voltage below 12 volts. A refractometer is used to test specific gravity.

Answer A is incorrect. A carbon pile tester is used to create a load on the alternator.

Answer B is incorrect. An amp clamp can be used to measure alternator output.

Answer D is incorrect. A voltmeter is used to measure alternator voltage.

Question #22

Answer C is correct. Both Technicians A and B are correct. It is important to check cylinder head bolts prior to installing them into the cylinder head, so Technician A is correct. Technician B is also correct. The cylinder block deck must be checked for warpage before installing the head. Readings beyond this specification indicate the need to resurface the block fire deck. However, you should always refer to the OEM service manual for the correct value. Blocks do not warp nearly as often as cylinder heads. When the block fire deck is warped or not parallel to the main bearing bores, it may be resurfaced with a milling machine or grinder. Next, check the main bearing bore and alignment. It is good shop practice to record the amount of stock removed from the deck by stamping the amount removed on the cylinder block pad.

Answer A is incorrect. Technician B is also correct.

Answer B is incorrect. Technician A is also correct.

Answer D is incorrect. Both technicians are correct.

Question #23

Answer B is correct. Only Technician B is correct. ELC should only have premix added, never water. Ethylene glycol is derived from ethylene oxide, which is produced from ethylene, a basic petroleum fraction. Propylene glycol is derived from propylene oxide, which is produced from propylene, another basic petroleum fraction. When mixed in a solution with water, both PG and EG are described as aqueous.

Answer A is incorrect. Propylene glycol- and ethylene glycol-based coolants should never be mixed. The mixture itself will not cause any engine or cooling system problems, but it will be impossible to determine the antifreeze mixture strength with either a refractometer or a hydrometer. If a mixture of EG and PG is known to have taken place and the coolant cannot be immediately replaced, use a refractometer with an EG and a PG scale, and average the two readings. However, the cooling system should be drained and refilled with either aqueous PG or EG when practical to avoid problems later on. ELC is only sold premixed and is dyed a red color.

Answer C is incorrect. Only Technician B is correct.

Answer D is incorrect. Only Technician B is correct.

Question #24

Answer A is correct. A plugged fuel tank breather can create a below-atmospheric pressure in the fuel tank that the fuel transfer pump is unable to overcome, resulting in engine shutdown due to fuel starvation. Currently, most jurisdictions in North America permit venting of diesel fuel tanks to the atmosphere. Therefore, as fuel is pumped out of on-board tanks, it is replaced by ambient air drawn in through vent valves.

Answer B is incorrect. A leaking intake manifold would result in a high whistling sound and possible dirt ingestion in the engine; however, it would not starve for fuel.

Answer C is incorrect. Insufficient throttle travel would cause low power, but not fuel starvation.

Answer C is incorrect. A missing fuel tank cap would cause contaminated fuel, but would not directly cause fuel starvation.

Question #25
Answer D is correct. Neither technician is correct. Oil pump regulating valves are both adjustable and rebuildable. Pressure-regulating valves are responsible for defining the maximum system oil pressure; most are adjustable. Typically, an oil pressure-regulating valve consists of a valve body with an inlet sealed with a spring-loaded, ball-check valve. Other types of poppet valves are also used, but the principle is the same. The regulating valve body is plumbed in parallel to the main oil pump discharge line. When oil pressure is sufficient to unseat the spring-loaded check ball, it unseals, permitting oil to pass through the valve and spill to the oil sump. This action will cause the pressure in the oil pump discharge line to drop.

Answer A is incorrect. Technician A is incorrect.

Answer B is incorrect. Technician B is incorrect.

Answer C is incorrect. Neither technician is correct.

Question #26
Answer D is correct. Neither technician is correct. Sleeves should be removed with a puller and adapter plate or shoe. The procedure is obviously more simple on wet liners. The tool shown is a liner puller.

Answer A is incorrect. Technician A is incorrect.

Answer B is incorrect. Technician B is incorrect.

Answer C is incorrect. Neither technician is correct.

Question #27
Answer C is correct. Both Technicians A and B are correct. Most engine management computers are equipped with a means of displaying fault codes without the use of diagnostic instruments. This can be by means of flashing codes as an example. Most current vehicles also now have digital dash displays that alert the driver to any conditions that affect the operation of the vehicle.

Answer A is incorrect. Technician B is also correct.

Answer B is incorrect. Technician A is also correct.

Answer D is incorrect. Both technicians are correct.

Question #28
Answer A is correct. Only Technician A is correct. One of the effects of too much clearance between the piston and the cylinder wall is piston rocking. If the pistons are allowed to rock, they will stress the sides and crack the skirts. Piston-to-cylinder clearance should be kept to specification to allow the pistons to remain straight in the bore.

Answer B is incorrect. Running the engine with a low oil level may cause piston and cylinder scuffing, but will not cause cracked skirts.

Answer C is incorrect. Only Technician A is correct.

Answer D is incorrect. Only Technician A is correct.

Question #29

Answer D is correct. Droplets of water or gray oil on a dipstick would LEAST-Likely be caused by a worn oil pump. All the other answers are indications that coolant has leaked into the oil. Coolant in the engine lubricant gives it a milky, cloudy appearance when churned into the oil. After settling, the coolant will usually collect at the bottom of the oil sump. When the drain plug is removed, the heavier coolant will exit first, as long as sufficient time has passed since running the engine. When coolant is found in the engine oil, the cylinder-head injector tubes and cylinder head gasket are the most likely sources.

Answer A is incorrect. A cracked block or cylinder head could cause this condition.

Answer B is incorrect. A blown head gasket could cause this condition.

Answer C is incorrect. A leaking oil cooler bundle could cause this condition.

Question #30

Answer A is correct. Only Technician A is correct. Low manifold boost can be caused by air inlet restriction so Technician A is correct.

Answer B is incorrect. Manifold boost cannot be checked with a water manometer. Peak boost in truck diesel engines is typically 35 psi, which is equivalent to 970 inches of water in a manometer.

Answer C is incorrect. Only Technician A is correct.

Answer D is incorrect. Only Technician A is correct.

Question #31

Answer C is correct. Both Technicians A and B are correct. Valve rotators use a ratchet principal or a ball and coaxial spring to rotate the valve fractionally each time it is actuated. Valve rotation should be checked after assembly by marking an edge of the stem and then tapping the valve stem with a light nylon hammer a number of times. The valve should visibly rotate each time the stem is struck. Technician A's method of verifying whether valve rotators are rotating is good practice so he is correct. Technician B is also correct in saying that defective valve seals can cause oil consumption. Oil will leak past defective valve guide seals and into the combustion chamber.

Answer A is incorrect. Technician B is also correct.

Answer B is incorrect. Technician A is also correct.

Answer D is incorrect. Both technicians are correct.

Question #32

Answer B is correct. Small leaks in the intake manifold gasket will cause boost pressure to build very slowly, affecting acceleration. Once the vehicle speed has increased, boost pressure is much higher and the reduced boost level is less noticeable. At idle speeds, little to no boost is produced, so the leak will not affect engine operation at idle speeds.

Answer A is incorrect. An inoperative wastegate actuator would have no affect on acceleration. The boost levels would be able to go higher than normal, possibly damaging air system components.

Answer C is incorrect. A clogged air filter would cause engine performance problems through all speed ranges.

Answer D is incorrect. A plugged EGR port would affect EGR system operation, but would have no affect on boost pressure.

Question #33

Answer A is correct. All the steps listed are required to maintain effective cooling in a diesel engine EXCEPT flushing and refilling of the cooling system annually. Some extended-life cooling systems are designed to last for up to five years and back-flushing is seldom recommended by diesel engine manufacturers.

Answer B is incorrect. Coolant should be tested regularly.

Answer C is incorrect. Air must be bleed from the cooling system during refilling to prevent engine damaging air pockets.

Answer D is incorrect. Coolant must be tested for antifreeze protection level. This is usually done by test strips or a refractometer.

Question #34
Answer B is correct. Only Technician B is correct. Fault codes in most diesel engine management systems are logged in EEPROM, so the act of disconnecting the vehicle batteries is not going to erase them.

Answer A is incorrect. Most modern diesel engine ECMs write fault code information to the nonvolatile memory, which is not erased when power is removed. Usually the only way to erase fault codes is through the use of approved test equipment, so Technician A is incorrect.

Answer C is incorrect. Only Technician B is correct.

Answer D is incorrect. Only Technician B is correct.

Question #35
Answer C is correct. Both Technicians A and B are correct. Extended Life Coolants (ELC) are low-maintenance coolants that may require only one SCA treatment in six years. The engine-cooling fan may be engaged for engine braking as part of engine brake strategy. ELCs promise a service life of 600,000 miles (960,000 km) or six years, with one additive recharge at 300,000 miles (480,000 km) or three years. This compares with a typical service life of two years during which up to 20 recharges of SCA would be required for conventional EG and PGs. ELCs are presently available only as a premixed solution to ensure that the water quality is at the required level. The pricing of ELCs by quantity is generally comparable with EG and PG, but because of reduced cooling system maintenance and extended service life, they will probably become the coolant of choice of the engine OEMs. No test kits are required to monitor the ELC SCA level.

Answer A is incorrect. Technician B is also correct.

Answer B is incorrect. Technician A is also correct.

Answer D is incorrect. Both technicians are correct.

Question #36
Answer A is correct. When troubleshooting a suspected blown cylinder head gasket, the first step should be to check the head bolt torque values. The other answers involve steps that would be performed after the cylinder head is removed.

Answer B is incorrect. Magna fluxing occurs after head removal.

Answer C is incorrect. Visual inspection occurs after removal.

Answer D is incorrect. Pressure testing occurs after removal.

Question #37
Answer C is correct. Both Technicians A and B are correct. ELCs are premixed and should not be diluted. Silicone hoses should have clamps designed for them and should be carefully torqued, since silicone hoses may be damaged by overtightening.

Answer A is incorrect. Technician B is also correct.

Answer B is incorrect. Technician A is also correct.

Answer D is incorrect. Both technicians are correct.

Question #38
Answer A is correct. Only Technician A is correct in describing how manufacturers want cylinder-head bolts torqued.

Answer B is incorrect. It is essential that bent push tubes be replaced rather than straightened. When straightened rods are returned to service, they fail quickly and can cause severe and expensive engine damage.

Answer C is incorrect. Only Technician A is correct.

Answer D is incorrect. Only Technician A is correct.

Question #39
Answer B is correct. Only Technician B is correct. The description perfectly describes what is often called "bottom-end" or "big-end" knock. This condition is caused by a failed connecting rod bearing.

Answer A is incorrect. Worn crankshaft main bearings are unlikely to cause the thumping noise described.

Answer C is incorrect. Only Technician B is correct.

Answer D is incorrect. Only Technician B is correct.

Question #40
Answer B is correct. The correct answer is B. Only Technician B is correct. Although the engine door wiring needs to be repaired, the cause of the damage must be found and corrected first. A common cause of engine door wiring damage is missing door stops or bumpers. If missing, the engine door may close too far and allow fan belts and pulleys to come in contact with the wiring, causing damage. If not found and corrected, the repaired wiring will only be damaged again when the vehicle is placed back into service.

Answer A is incorrect. The cause of the damage must also be found and corrected.

Answer C is incorrect. Only Technician B is correct.

Answer D is incorrect. Only Technician B is correct.

Question #41
Answer A is correct. Only Technician A is correct. On mechanically controlled engines, the throttle lever controls the engine's speed. Actuation may be by linkage, cables, or air-operated actuators. The throttle return to idle speed position is controlled by a return spring. If the spring becomes broken or weak, the throttle may not return to idle speed, causing a sticking throttle condition.

Answer B is incorrect. Aerated fuel will cause engine performance problems, but will not cause RPMs to hang above idle speed.

Answer C is incorrect. Only Technician A is correct.

Answer D is incorrect. Only Technician A is correct.

Question #42
Answer C is correct. Both Technicians A and B are correct. Customer data programming can be changed using a PC and proprietary software, though some or all fields may be password protected. Changing customer data programming options can result in less efficient engine performance because, depending on the system, engine parameters such as peak torque, torque rise, rated power rpm, and rated power itself can be changed.

Answer A is incorrect. Technician B is also correct.

Answer B is incorrect. Technician A is also correct.

Answer D is incorrect. Both technicians are correct.

Question #43
Answer C is correct. Both Technicians A and B are correct. Technician A is describing a correct method of aligning cylinder heads on a multicylinder head engine. Technician B is also correct because bolt threads should be free of dirt, which could bind in the threads, and free of and oil or other fluids, which could cause hydrostatic lock. On in-line multicylinder heads, separate heads

must be aligned with a straightedge across the intake manifold faces before torquing. Failure to observe torquing increments and sequencing can result in cracked cylinder heads, failed head gaskets, and fire rings that will not seal. Because of the large number of fasteners involved, a click-type torque wrench should be used. Some OEMs require that a torque-turn be used. This requires setting a torque value first and then turning a set number of degrees beyond that value using a template or protractor. Cylinder head bolts should be lubricated according to manufacturer specifications, although excessive quantities of oil should be avoided because the excess can drain into the bolt hole and cause a hydraulic lock.

Answer A is incorrect. Technician B is also correct.

Answer B is incorrect. Technician A is also correct.

Answer D is incorrect. Both technicians are correct.

Question #44
Answer C is correct. Both Technicians A and B are correct. For each 1 psi that a liquid is pressurized above atmospheric pressure, its boiling point raises 3°F at sea level. Most diesel engine cooling systems maintain a system pressure of 10-20 psi through the use of a relief valve in the radiator surge tank, so Technician A is correct. Antifreeze, when mixed with water, lowers the freeze point and raises the boil point so Technician B is also right.

Answer A is incorrect. Technician B is also correct.

Answer B is incorrect. Technician A is also correct.

Answer D is incorrect. Both technicians are correct.

Question #45
Answer C is correct. Both Technicians A and B are correct. A test gauge is used to verify readings produced by the dash panel instrumentation. Technician A is correct. The readings for both the dash and master gauges should be identical. If not, the operation of the dash gauge/signal/sensor circuit will have to be checked. Technician B is also correct in saying that test readings should be taken through a range of operating rpms.

Answer A is incorrect. Technician B is also correct.

Answer B is incorrect. Technician A is also correct.

Answer D is incorrect. Both technicians are correct.

Question #46
Answer A is correct. The piston illustrated was subject to excessive cylinder temperatures. Because all the other pistons did not show this wear, the problem must be a single cylinder issue. A worn exhaust lobe would prevent the exhaust gasses from escaping the cylinder on the exhaust stroke, thus creating excessive cylinder temperatures in a single cylinder.

Answer B is incorrect. A restricted exhaust system could cause excessive cylinder temperatures; however, all cylinders would be affected.

Answer C is incorrect. A restricted air cleaner could also cause excessive cylinder temperatures; however, it would affect all cylinders.

Answer D is incorrect. A worn injector lobe would affect only one cylinder, but it would not cause high temperature. It would cause cooler temperatures because little or no fuel would be injected.

Question #47
Answer D is correct. Neither technician is correct. When installing silicone gaskets into a rocker housing cover, the manufacturer service instructions should be observed.

Answer A is incorrect. Technician A is incorrect because installing silicone gaskets into a rocker housing cover will seldom involve applying adhesive to the gasket.

Answer B is incorrect. Technician B is incorrect. Many rocker housing gaskets on modern engines are designed to be reused and their price reflects this.

Answer C is incorrect. Neither technician is correct.

Question #48
Answer A is correct. The fuel pressure sensor would *not* be part of the shutdown solenoid circuit.

Answer A is incorrect. A fuel pressure sensor can be part of the circuit.

Answer B is incorrect. The key switch can be part of the circuit.

Answer D is incorrect. The solenoid coil can be part of the circuit.

Question #49
Answer B is correct. If the fuel cut-off solenoid is not actuating, the fuel pump/injectors will not receive any fuel and the engine will not start.

Answer A is incorrect. If the air filter was clogged enough to prevent the engine from starting, it would not run when ether was used.

Answer C is incorrect. Low oil level will not prevent the engine from starting.

Answer B is incorrect. If the starter was defective, the engine would not even crank.

Question #50
Answer A is correct. To check flywheel runout, mount a dial indicator with a magnetic base on the flywheel housing, and check the flywheel runout at the clutch contact face. The typical allowable runout is .001 in. per inch clutch radius (0.001 mm per millimeter of clutch radius). Measure the radius from the center of the flywheel to the outer edge of the clutch contact face. For example, a 14-in. (355 mm) clutch would have a 7-in. (177 mm) radius and allow a 0.007-in. (0.177 mm) total indicator reading of the dial indicator. If the measured values fall out of these specifications, repeat the assembly process and remeasure until a cause can be determined. In the figure, the dial indicator is mounted to measure the clutch face surface runout, so the clutch bolts to the flywheel.

Answer B is incorrect. The pilot bore is not being measured.

Answer C is incorrect. Crankshaft radial runout is not being checked.

Answer D is incorrect. This is not the flywheel housing concentricity check.

Question #51
Answer D is correct. Neither technician is correct. If the EGR cooler were clogged, the only symptom would be little or no EGR flow into the intake. The engine would not overheat, but emissions would be affected. If the EGR valve were leaking, causing EGR flow all the time, the most noticeable symptoms would be reduced boost pressure, a decrease in engine performance, and possibly rough running, but the engine should not overheat.

Answer A is incorrect. Technician A is incorrect.

Answer B is incorrect. Technician B is incorrect.

Answer C is incorrect. Neither technician is correct.

Question #52
Answer D is correct. Although proper gear timing is essential to proper engine operation, incorrect timing mark alignment would not create excessive backlash. If gear timing marks are not aligned, excessive engine damage can result from valve to piston contact.

Answer A is incorrect. Worn idler gear bearings can cause excessive backlash.

Answer B is incorrect. Worn gear teeth can cause excessive backlash.

Answer C is incorrect. Worn camshaft bearings can cause excessive backlash.

Question #53

Answer C is correct. The clearance should be measured between the ring and the side of the piston ring groove with the appropriate thickness feeler gauge. If the clearance is too great, ring and land damage can result. If the clearance is too little, the rings may not fit into the grooves properly, preventing piston installation.

Answer A is incorrect. Plastigage is used to measure bearing to journal clearances.

Answer B is incorrect. Micrometers are used to measure thickness, not clearance between parts.

Answer D is incorrect. Dial bore gauges are used to measure cylinder diameters.

Question #54

Answer D is correct. The voltage drop test described is used to check the condition of the connection between the battery post and positive terminal so this is not a battery test.

Answer A is incorrect. A multimeter can be used to determine state of charge.

Answer B is incorrect. A carbon pile load tester can be used to test the ability of a battery to deliver sufficient amperage.

Answer C is incorrect. A refractometer can be used to test a battery.

Question #55

Answer A is correct. Only Technician A is correct. Oil trapped in blind bolt holes can cause hydrostatic lock when the bolt is installed and invalidate torque readings.

Answer B is incorrect. A cylinder head bolt with any visible damage should be replaced not repaired.

Answer C is incorrect. Only Technician A is correct.

Answer D is incorrect. Only Technician A is correct.

Question #56

Answer B is correct. In the figure, the dial indicator is located to check turbine shaft endplay.

Answer A is incorrect. Radial runout is checked by installing the dial indicator through the oil feed or oil drain hole and moving the shaft up and down.

Answer C is incorrect. Turbine fin wear is checked with a visual inspection.

Answer D is incorrect. Compressor fin wear is checked by visual inspection.

Question #57

Answer B is correct. The Most-Likely cause of this problem would be a leak at the primary filter housing allowing air to enter the fuel system.

Answer A is incorrect. A single seized injector nozzle would not cause the transfer pump to fail to build pressure and it would not prevent the engine from starting.

Answer C is incorrect. A restriction in the high-pressure line between the injector pump and injector would cause a single cylinder problem.

Answer D is incorrect. Restricted lines after the prime pump may cause a no-start condition; however, the hand-primer pump would build pressure.

Question #58

Answer A is correct. Only Technician A is correct. A diesel engine should never be cranked for a period exceeding 30 seconds for any reason.

Answer B is incorrect. Technician B is incorrect. A starter load test should be performed using the chassis batteries without any kind of external assist.

Answer C is incorrect. Only Technician A is correct.

Answer D is incorrect. Only Technician A is correct.

Question #59

Answer C is correct. Both Technicians A and B are correct. Checking the air intake system for restriction is a common method of assessing the serviceability of the air filter in that high readings will indicate a plugged filter, so Technician A is correct. Technician B's statement about taking the readings at high full load is correct. Maximum inlet restriction should be measured at rated speed with the engine fully loaded.

Answer A is incorrect. Technician B is also correct.

Answer B is incorrect. Technician A is also correct.

Answer D is incorrect. Both technicians are correct.

Question #60

Answer D is correct. The injector nozzle would not need to be replaced. So it is the LEAST-Likely to be required.

Answer A is incorrect. Leaking sleeves must be replaced.

Answer B is incorrect. Leaking injector sleeves (tubes) are a relatively common failure. In most cases, engine coolant comes into direct contact with the sleeve so coolant leakage problems result. When injector sleeves are found to be leaking, the sleeve should be replaced, the cylinder head should then be hot pressure tested.

Answer C is incorrect. Because nozzle-tip protrusion is dependent on the precise location of the injector sleeve, this should also be measured. The injector nozzle has nothing to do with the reasons for replacing the cylinder sleeve so this is the least item to be replaced.

Question #61

Answer C is correct. The figure shows a thermatic fan, one in which effective cycle is managed by a bimetal strip connected to a valve, which in turn controls silicone media to manage lock-up and free-wheel cycles.

Answer A is incorrect. This is not an air clutch.

Answer B is incorrect. This is not an electric clutch.

Answer D is incorrect. This is not a hydraulic clutch.

Question #62

Answer C is correct. A defective shutdown solenoid can cause no fuel delivery to the engine, resulting in a no-start condition. Failure of the fuel shutdown solenoid is a relatively common problem in distributor pumps, allowing the engine to crank but not receive any fuel.

Answer A is incorrect. One defective injector will usually not prevent a multicylinder engine from starting.

Answer B is incorrect. This engine cranks so the starter cut-out relay is not the cause of the problem.

Answer D is incorrect. A leaking air intake hose would not cause a no-start condition.

Question #63

Answer C is correct. Both Technicians A and B are correct. A cooling system's boil point is increased by holding it under pressure. If the pressure cannot be maintained, the coolant can boil at a lower temperature. Boiling coolant within a cooling system accelerates an overheat condition because a gas does not conduct heat as well as a liquid The radiator cap is designed to relieve system pressure when it exceeds its rating.

Answer A is incorrect. Technician B is also correct.

Answer B is incorrect. Technician A is also correct.

Answer D is incorrect. Both technicians are correct.

Question #64

Answer B is correct. The probable source of soot in the intake manifold of a turbocharged, charge air-cooled diesel is the turbocharger itself. The condition is usually caused by passing oil getting heated in the hot, turbo-boosted air. Some soot in the intake manifold is normal in most diesel engines; it usually indicates an engine operated at low loads and speeds for prolonged periods. In such operating conditions, the valve overlap duration is at a maximum in real-time values and manifold boost is minimal or nonexistent. Excessive soot in the intake manifold may indicate imminent turbocharger failure or an injection timing problem.

Answer A is incorrect. The intercooler cannot cause soot in the intake manifold.

Answer C is incorrect. The intake manifold cannot cause soot inside itself.

Answer D is incorrect. The head gasket could not cause soot to build up in the intake manifold.

Question #65

Answer B is correct. When a major diesel engine overhaul is performed, it is good practice to thoroughly inspect the cylinder block using magnetic flux inspection.

Answer A is incorrect. The painting would take place after it had passed magnetic flux crack detection.

Answer C is incorrect. The cylinder sleeves would be removed before cleaning.

Answer D is incorrect. Plug installation would take place after magna fluxing. It makes sense to magnetic flux test engine cylinder blocks, crankshafts, and all connecting rods at every major engine overhaul. These processes are neither expensive nor time consuming and as the consequences of a single warranted engine failure out of 20 overhauls will demolish the profits of the other 19, it is shortsighted to overlook it.

Question #66

Answer D is correct. When the by-pass valve trips on an oil filter pad assembly, the oil is routed around the filter assembly; that is, it bypasses it. This will only occur when the filter element has become plugged. Filter-mounting pad by-pass valves operate in much the same way as the oil pressure regulating valves, except that their objective is to route the lubricant around a restricted full-flow filter to prevent engine damage by oil starvation. When a filter by-pass valve is actuated and the check valve unseated, instead of spilling the oil to the crankcase, it reroutes the oil directly to the lubrication circuit, effectively shorting out the filter assembly. So, when a by-pass value trips, unfiltered oil is circulated through the lubrication circuit.

Answer A is incorrect. A restricted inlet would cause insufficient oil to the filter.

Answer B is incorrect. If the oil pump was sucking air, pressure would be low.

Answer C is incorrect. A massive oil leak would lower pressure, not raise it.

Question #67

Answer A is correct. Only Technician A is correct. When an engine shows wear only in a specific portion and the rest of the engine looks normal, the technician should concentrate the diagnostic routine on items that affect that part of the engine only. At first glance, both items mentioned would seem to affect the entire engine; therefore, they are not very likely to cause damage to only the rear two cylinders. If a diesel engine was operated with dirty oil, then all the bearing surfaces should be worn. All cylinders would be affected if the water pump produced inadequate flow, but the rear two cylinders would be the most affected due to distance from the pump. Technician A is correct.

Answer B is incorrect. Dirty oil would most likely affect the entire engine.

Answer C is incorrect. Only Technician A is correct.

Answer D is incorrect. Only Technician A is correct.

Question #68

Answer B is correct. High manifold boost increases the air density and the potential to produce high power so this would be the LEAST-Likely cause of a lack of power complaint.

Answer A is incorrect. High exhaust back pressure would cause low power and overheating.

Answer C is incorrect. Clogged cooling fines would cause high air inlet temperatures that would reduce horsepower.

Answer D is incorrect. A small leak in the suction fuel line would cause low power.

Question #69

Answer C is correct. Both Technicians A and B are correct. Worn engine lube pumps can cause low oil pressure. HEUI high-pressure oil pumps do use a swash plate (wobble plate) principle.

Answer A is incorrect. Technician B is also correct.

Answer B is incorrect. Technician A is also correct.

Answer D is incorrect. Both technicians are correct.

Question #70

Answer D is correct. Neither technician is correct. Bearing shells should be installed to the bore clean and dry to optimize heat transfer between the bearing material and the bore. A thin coat of oil may be applied to the bearing face only. When Plastigage is used to check bearing clearance, it is important that the crankshaft not be rotated. Plastigage is soft plastic thread that easily conforms to whatever clearance space is available when compressed between a bearing and journal. The crushed width can then be measured against a scale on the Plastigage packaging. The less clearance available will result in the Plastigage being flattened to a wider dimension. A short strip of Plastigage should be cut and placed across the center of the bearing in line with the crankshaft. The bearing cap with the bearing shell in place should then be installed and torqued to specification in the incremental steps outlined in the workshop manual. Do not rotate the engine with the Plastigage in place. Next, the bearing cap and shell should be removed and the width of the flattened Plastigage checked against the dimensional gauge on the packaging. If clearance is within specifications, carefully remove the Plastigage from the journal before reinstalling the cap and shell assembly.

Answer A is incorrect. Technician A is incorrect.

Answer B is incorrect. Technician B is incorrect.

Answer C is incorrect. Neither technician is correct.

Question #71

Answer C is correct. Both Technicians A and B are correct. If necessary, purge the high-pressure fuel lines of air by loosening the connector one-half to one turn and cranking the engine until fuel, free from bubbles, sprays from the connection. Checking for air being pulled into the fuel subsystem is a good method as this is especially important when working with fuel subsystems that are entirely under suction.

Answer A is incorrect. Technician B is also correct.

Answer B is incorrect. Technician A is also correct.

Answer D is incorrect. Both technicians are correct.

Question #72

Answer B is correct. The figure shows a dry-type, positive filtration, air filter assembly. This type of filter has high efficiencies because all of the air has to pass through the filtering media to get into the intake system. Some dry element filters are two-stage and may eliminate the need for a precleaner; these induce a vortex flow and use centrifugal force to separate heavier particulate, which is then discharged by an ejector valve.

Answer A is incorrect. This is not an oil bath air cleaner.

Answer C is incorrect. This is not a centrifugal air cleaner.

Answer D is incorrect. This is not a precleaner.

Question #73
Answer D is correct. Neither technician is correct.

Answer A is incorrect. Heads are checked for warpage with a straightedge and feeler gauge.

Answer B is incorrect. Wrist pins are checked for wear with an outside micrometer.

Answer C is incorrect. Neither technician is correct.

Question #74
Answer A is correct. Only Technician A is correct. One lead of the voltmeter is connected to the ground terminal of the battery and the other lead is connected to the base of the starter. Then the voltmeter is read while the engine is cranked. If the voltage reading exceeds 0.2 volt, excessive resistance exists in the ground circuit.

Answer B is incorrect. A defective starter drive is a mechanical problem that would not affect the voltage at the starter.

Answer C is incorrect. Only Technician A is correct.

Answer D is incorrect. Only Technician A is correct.

Question #75
Answer A is correct. Only Technician A is correct. The DPF needs to be serviced. DPFs are designed to trap and collect solid matter, such as soot and ash, present in the exhaust. As these solids accumulate, the exhaust flow through the DPF will be reduced, creating high back pressure and temperature levels. When the DPF reaches this point, it must be either removed and cleaned, following manufacturer's procedures, or replaced.

Answer B is incorrect. While a burned exhaust valve can cause elevated exhaust temperatures by allowing unburned fuel to enter the catalyst assembly and generate high internal temperatures, it would not cause an increase in back pressure.

Answer C is incorrect. Only Technician A is correct.

Answer D is incorrect. Only Technician A is correct.

Question #76
Answer D is correct. Neither technician is correct.

Answer A is incorrect. While some manufacturers recommend replacing the damper during engine rebuilds, some manufacturers allow reuse after careful inspection for damage. Service manuals should be consulted for determination of reuse.

Answer B is incorrect. No manufacturer will allow repairing of the damper, especially welding of cracks. The job of the damper is to remove vibration and help balance the crankshaft, and welding could add an imbalance to crankshaft, causing premature failure of the crankshaft.

Answer C is incorrect. Neither technician is correct.

Question #77
Answer C is correct. An electronically controlled diesel engine should set active codes for systems or sensors that are currently operating outside of preprogrammed parameters. An example would be an out-of-range coolant temperature sensor. If several codes are all set at the same time, it is highly likely there is a common cause. In this situation the most reasonable diagnostic course would be to check for adequate power and ground to the ECM.

Answer A is incorrect. A worn crankshaft thrust washer would most likely set only a crankshaft position sensor code.

Answer B is incorrect. A worn camshaft thrust washer would most likely only set a camshaft position sensor code.

Answer D is incorrect. It is highly unlikely the ECM is faulty.

Question #78
Answer D is correct. Neither technician is correct. Damaged camshaft bearing surfaces should not be turned and diesel engine camshafts are never repaired with silver solder. The technician should be aware that the smallest indication of a hard-surfacing failure on the cam profile requires little time to advance to a total failure.

Answer A is incorrect. Technician A is incorrect.

Answer B is incorrect. Technician B is incorrect.

Answer C is incorrect. Neither technician is correct.

Question #79
Answer A is correct. Before installing valves into a cylinder head, seat contact should be tested with Prussian blue. Check also for the correct pattern on the valve face. The remainder of the procedures would be performed after the cylinder valves had been installed in the head.

Answer B is incorrect. Rocker arms are adjusted after the head is installed.

Answer C is incorrect. Measuring camshaft lift is checked while inspecting the camshaft.

Answer D is incorrect. Head gaskets are checked during disassembly for evidence of leaks. Head gaskets are replaced. They are not reused.

Question #80
Answer D is correct. While some engine manufacturers require that gear installation be performed before the cam is installed, some do not. Either way, the presence of the gear will have no affect on the bearing clearance or alignment.

Answer A is incorrect. If the camshaft is bent, it will bind in the bore and cause it to lock.

Answer B is incorrect. If the camshaft bearing bores are not in-line, the cam will bind and be hard to turn.

Answer C is incorrect. If there is insufficient clearance, the cam will bind and be hard to turn. This condition will also cause bearing and camshaft damage.

Question #81
Answer B is correct. Only Technician B is correct. If a computer-controlled diesel engine will start but not accelerate, the most likely problem is a fault in the throttle position sensor circuit. The ECM does not recognize that the operator wants to increase engine speed.

Answer A is incorrect. The intake air temperature sensor helps the computer trim the fuel quantity and timing to produce the cleanest burn under all incoming air temperatures.

Answer C is incorrect. Only Technician B is correct.

Answer D is incorrect. Only Technician B is correct.

Question #82
Answer D is correct. Minimum head "deck-to-deck" thickness is critical. If a cylinder head is machined below this minimum specification, it will be much more likely to crack due to its now thinner casting. Additionally, when the head becomes thinner, critical valvetrain geometry angles are changed. The figure indicates how much machining can be sustained while meeting the minimum head deck-to-deck specification.

Answer A is incorrect. This measurement does not prevent head warpage; however, if a head is too thin it will be prone to warpage.

Answer B is incorrect. Valve guide height is measured on the valve guide.

Answer C is incorrect. Cylinder head warpage is determined with a straightedge and feeler gauge.

Question #83
Answer C is correct. It is bad practice to use a test light on electronic circuits, so this should be the LEAST-Likely. A test light can draw too much amperage.

Answer A is incorrect. A digital ohmmeter is a good tool to use on an electronic circuit.

Answer B is incorrect. A digital multimeter is a good tool to use on an electronic circuit. DMMs are simple tools for taking electrical measurements. DMMs may have any number of special features but essentially, they measure electrical pressure or volts, electrical current flow or amps, and electrical resistance or ohms. A good-quality DMM with minimal features may be purchased for as little as $100; as the features, resolution, and display quality increase in sophistication, the price increases proportionally. Because most electronic circuit testing requires the use of a DMM, this instrument should replace the analog multimeter and circuit test light in the truck/bus technician's toolbox. Reliability, accuracy, and ease of use are all factors that should be considered when selecting a DMM for purchase. Some options the technician may want to consider are a protective rubber holster (will greatly extend the life of the instrument!), analog bar graphs, and enhanced resolution.

Answer D is incorrect. A breakout box is a good tool to use on an electronic circuit.

Question #84
Answer A is correct. LS (limiting speed) governing means that accelerator pedal angle is correlated to fuel quantity. If more fuel is required, for instance to climb a hill, the pedal angle must be increased. This action/response is similar to the accelerator pedal in a car.

Answer B is incorrect. A variable speed (VS) governor is known as an all-speed governor in the United Kingdom. It sets engine idle speed and defines high idle and any speed in the intermediate range, depending on accelerator pedal position.

Answer C is incorrect.

Answer D is incorrect.

Question #85
Answer D is correct. Neither technician is correct. Audit trails are designed to provide a profile of vehicle handling data for later analysis and cannot be erased in the same way as historic fault, so Technician A is incorrect. Technician B is also incorrect. Audit trails are not written to PROM, they are logged in EEPROM. Also, because of EEPROM capability, PROM chips are not replaceable in current diesel engine management systems.

Answer A is incorrect. Technician A is incorrect.

Answer B is incorrect. Technician B is incorrect.

Answer C is incorrect. Neither technician is correct.

Question #86
Answer C is correct. Both Technicians A and B are correct. Diesel engine manufacturers do not recommend any machining procedures on camshafts. The camshaft gear must be visually inspected to assess its serviceability. Visual inspection should be sufficient to determine the condition of timing gears. Any indications of cracks, pitting, heat discoloration, or chipping of the gear teeth will predicate its replacement. Press-fit gears will require the use of a mechanical, pneumatic, or hydraulic puller to remove the gear from the shaft.

Answer A is incorrect. Technician B is also correct.

Answer B is incorrect. Technician A is also correct.

Answer D is incorrect. Both technicians are correct.

Question #87

Answer B is correct. When replacing connectors, ensure that each wire is labeled by cavity location to facilitate reassembly. Use strips of masking tape on wires to write each cavity code on. This is usually easier than reassembling a multiwire connector using the OEM wiring schematic. The figure shows a Weather-pack type connector into which a release tool is being inserted to release the lock tang.

Answer A is incorrect. This is not a voltage drop test.

Answer C is incorrect. The terminal blades are not being spread; the locking tang is being released.

Answer D is incorrect. The figure does not show a digital multimeter test probe being inserted; it shows a release tool being inserted. A digital multimeter test probe should not be inserted into the cavity as it will most likely spread the cavity open, resulting in a poor electrical connection when the mating connector is reconnected.

Question #88

Answer D is correct. When installing rocker arms, the bushings, shaft, and pallet (actuation end) must be inspected. However, the adjustment screws are not adjusted until the valve lash is set, which happens after the assembly.

Answer A is incorrect. Rocker arms are checked prior to installing.

Answer B is incorrect. The rocker shaft is inspected prior to installation.

Answer C is incorrect. The pallet is inspected prior to reassembly.

Question #89

Answer A is correct. Only Technician A is correct. Most bus TPSs are connected to V-Ref, usually 5 VDC. (Remember, one of the major engine manufacturers powers up its TPS with 8 VDC to output a PWM signal.)

Answer B is incorrect. The TPS is a potentiometer or voltage divider, so only a portion of V-Ref is returned as a signal. That would result in values always less than 5 VDC. The critical sensor in the electronic foot pedal assembly (EFPA) is the throttle position sensor (TPS). The TPS receives reference voltage (V-Ref) of 5 V and returns a portion of it, proportional to pedal mechanical travel.

Answer C is incorrect. Only Technician A is correct.

Answer D is incorrect. Only Technician A is correct.

Question #90

Answer C is correct. Both Technicians A and B are correct. Technician A correctly describes the operation of a thermistor. Technician B is also correct in saying that NTC thermistors are generally used in diesel engines.

Answer A is incorrect. Technician B is also correct.

Answer B is incorrect. Technician A is also correct.

Answer D is incorrect. Both technicians are correct.

Question #91

Answer B is correct. Oil with a high viscosity index changes little over a wide temperature range and therefore would be very unlikely to cause low oil pressure.

Answer A is incorrect. Worn bearings are a very common cause of low oil pressure.

Answer C is incorrect. A restricted oil pump suction tube would cause low oil pressure, especially at higher rpms when oil flow is at the highest.

Answer D is incorrect. A worn oil pump would bypass oil internally and cause low oil pressure.

Question #92

Answer C is correct. Both Technicians A and B are correct. A common cause of gray or black smoke is air starvation, indicating that an insufficient amount of air is in the engine cylinder to combust the fuel injected, so Technician A is correct. Overfueling can also cause gray or black smoke emission for the same reasons, so Technician B is also right.

Answer A is incorrect. Technician B is also correct.

Answer B is incorrect. Technician A is also correct.

Answer D is incorrect. Both technicians are correct.

Question #93

Answer B is correct. Only Technician B is correct. You must turn off the service valves before removing the old filter, otherwise, coolant will leak from the filter housing once the filter is removed. Turning off the valves isolates the filter housing from the rest of the cooling system.

Answer A is incorrect. Most manufacturers do not recommend priming the filter before installation. Any air in the filter will be purged once the engine is started and the coolant is allowed to circulate through the filter. Coolant level should be checked after servicing the filter.

Answer C is incorrect. Only Technician B is correct.

Answer D is incorrect. Only Technician B is correct.

Question #94

Answer B is correct. If both batteries are 12 volts and they supply a 12/12 electrical system, the batteries must be connected in parallel.

Answer A is incorrect. A series circuit would produce 24 volts.

Answer C is incorrect. A series/parallel circuit would be used with series/parallel magnetic switch in a 12/24 circuit.

Answer D is incorrect. An isolation circuit typically isolates a ground on a 12/24 circuit.

Question #95

Answer C is correct. Both Technicians A and B are correct. Whenever a crankshaft is damaged, the cylinder block should be checked. The master bar is designed to check the main align bore; it is installed without bearing shells with the main caps snug. If it does not rotate freely, the block align bore requires machining. The main bores should also be checked for out-of-round and this can be done with an inside micrometer.

Answer A is incorrect. Technician B is also correct.

Answer B is incorrect. Technician A is also correct.

Answer D is incorrect. Both technicians are correct.

Question #96

Answer A is correct. The figure shows a technician replacing valve guide seals.

Answer B is incorrect. The figure shows a technician replacing valve guide seals.

Answer C is incorrect. The figure shows a technician replacing valve guide seals.

Answer D is incorrect. The figure shows a technician replacing valve guide seals.

Question #97

Answer B is correct. Only Technician B is correct. Each lifter should be checked in its mating bore. Lifters are manufactured from cast iron and middle alloy steels and are usually located in guide bores in the cylinder block, allowing them to ride the cam profile over which they are positioned; pushrods are fitted to lifter sockets. The critical surface of a solid lifter is the face that directly contacts the cam profile. This face must be durable and may either be chemically hardened, cladded with a toughened alloy, or have a disc of special alloy steel molecularly bonded to the face. Solid lifters should be carefully examined primarily for thrust face wear at engine overhaul but also stem and socket wear. Guide bores in the cylinder block should also be measured using digital calipers or a telescoping gauge and a micrometer. Sleeving lifter guide bores is a relatively simple procedure that involves boring to an interference fit to the new sleeve's outside diameter. Always check that the lifters do not drag or cock in newly sleeved guide bores.

Answer A is incorrect. There is no need to replace the timing gears simply because you replaced the lifters.

Answer C is incorrect. Only Technician B is correct.

Answer D is incorrect. Only Technician B is correct.

Question #98

Answer B is correct. A voltage drop test is performed by placing the voltmeter leads across the portion of the circuit being tested and observing voltage with the circuit operating. In the figure, one lead is on the battery negative post; the other is attached to the starter case. The groundside of the circuit is between the voltmeter leads. The starter motor must be cranking when performing this test.

Answer A is incorrect. This is a voltage drop test on the ground side of the starter.

Answer C is incorrect. This is a voltage drop test on the ground side of the starter.

Answer D is incorrect. This is a voltage drop test on the ground side of the starter.

Question #99

Answer B is correct. Only Technician B is correct. Factory service bulletins are the most up-to-date information available for service issues. Printed service manuals may be out of date or contain errors that would be addressed by a service bulletin.

Answer A is incorrect. If a procedure or specification has been changed by the manufacturer after the publish date of the service manual, the information is no longer valid. Checks for updates should be made on a regular basis.

Answer C is incorrect. Only Technician B is correct.

Answer D is incorrect. Only Technician B is correct.

Question #100

Answer B is correct. The figures with this question show some typical torque sequence templates that the technician should observe when torquing cylinder heads. You will notice that the sequence starts in the middle to provide an even distribution of clamping force. This illustrates how the clamping force moves out from the middle when you tighten a cylinder head. The objective of torquing cylinder head gaskets is to ensure that the required amount of clamping force is obtained and the gasket conforms to its engineered yield shape. Some OEMs use the torque-turn method of clamping down the cylinder head. You initially torque the head to a small increment, such as 20 ft.-lb., then follow this torque up with a 90° (one-quarter) turn.

Answer A is incorrect. This numbering shows torque sequence.

Answer C is incorrect. The serial number is stamped into the casting.

Answer D is incorrect. The part number is cast into the head.

Question #101

Answer C is correct. Both Technicians A and B are correct. Some throttle position sensors are designed to output a digital PWM signal to the ECM. TPS calibration can be guided by the manufacturer software so that mechanical pedal travel can be phased with the signal produced.

Answer A is incorrect. Technician B is also correct.

Answer B is incorrect. Technician A is also correct.

Answer D is incorrect. Both technicians are correct.

Question #102

Answer B is correct. Only Technician B is correct. Testing an oil cooler bundle should be undertaken at a major engine overhaul and would not be a part of routine PM scheduling.

Answer A is incorrect. The oil cooler bundle does not need to be tested at every PMI.

Answer C is incorrect. Only Technician B is correct.

Answer D is incorrect. Only Technician B is correct.

Question #103

Answer C is correct. The figure shows the flywheel housing being checked for radial runout. This is a critical measurement that should be performed whenever a flywheel housing is removed and reinstalled from the engine block. If this measurement is out of specification, the result can be crankshaft damage, engine mount damage, flywheel housing damage, and transmission failures.

Answer A is incorrect. This is not the lateral runout test.

Answer B is incorrect. The dial indicator is not on the housing face.

Answer D is incorrect. The dial indicator is not on the pilot bore.

Question #104

Answer A is correct. Only Technician A is correct. Some possible causes of high pyrometer readings are: air inlet restriction, flow restricted, boost air heat exchanger, high engine load, fuel injection timing, and overfueling.

Answer B is incorrect. Air starvation does not result in lower cylinder temperatures. In fact, the temperatures would be higher, not lower.

Answer C is incorrect. Only Technician A is correct.

Answer D is incorrect. Only Technician A is correct.

Question #105

Answer D is correct. A mechanical variable-speed governor has no way of keeping the operator from lugging the engine. It sets engine idle speed and defines high idle and any speed in the intermediate range, depending on accelerator pedal position. A given amount of accelerator pedal travel will correspond to an engine rotational speed. As engine loading either increases or decreases, the governor will manage fueling to attempt to maintain that engine speed. Hydromechanical variable speed governors were common in many Mack Trucks and Caterpillar applications, among others, where power take-off (PTO) management was a consideration. PTO management is used when the engine drives auxiliary equipment. From the driver perspective, the VS governor takes a little getting used to. Most of today's electronic management systems can be toggled to either LS or VS mode. A governor classified as variable speed will usually provide excess startup fuel, define a torque rise profile, define droop curve, and be capable of no-fueling the engine for shutdown.

Answer A is incorrect. A mechanical variable speed governor controls idle speed.

Answer B is incorrect. A mechanical variable speed governor does meter fuel to the injectors.

Answer C is incorrect. A mechanical variable speed governor does control maximum engine speed.

Question #106

Answer C is correct. Both Technicians A and B are correct. Dash gauges are a better indicator of imminent problems than dash warning lights because they are capable of monitoring a worsening condition, giving the driver some advanced warning of a failure. Gauge failures can be caused by conditions other than the gauge circuit.

Answer A is incorrect. Technician B is also correct.

Answer B is incorrect. Technician A is also correct.

Answer D is incorrect. Both technicians are correct.

Question #107
Answer C is correct. Both Technicians A and B are correct. Technician A correctly describes the actuation medium in a HEUI injector. Also, HEUI injectors are capable of rate shaping; that is, precisely defining the speed at which the fuel injection plunger is driven into the pump chamber by managing the actuation oil pressure.

Answer A is incorrect. Technician B is also correct.

Answer B is incorrect. Technician A is also correct.

Answer D is incorrect. Both technicians are correct.

Question #108
Answer C is correct. Before a load test is performed on a battery, some specific criteria have to be met, including having the battery properly charged. If the battery then fails the load test, it should be replaced. The load, or capacity, test determines how well any type of battery, sealed or unsealed, functions under a load.

Answer A is incorrect. If the test was performed correctly, then the battery was already adequately charged.

Answer B is incorrect. The battery was fully charged before the test, therefore charging the battery a second time will not change the results of the test.

Answer D is incorrect. There is no reason to believe the voltage regulator is faulty; even if it is, replacing the regulator will not fix a faulty battery.

Question #109
Answer A is correct. Only Technician A is correct. Corrosion on wiring connectors on computer-controlled engines can affect engine performance and cause codes to be logged.

Answer B is incorrect. Many terminals on vehicle electronic system wiring harnesses are manufactured out of materials such as gold and platinum to provide reliable continuity and corrosion resistance. Unless the manufacturer recommends treatment of terminals and connectors with a corrosion inhibitor, it should not be used.

Answer C is incorrect. Only Technician A is correct.

Answer D is incorrect. Only Technician A is correct.

Question #110
Answer A is correct. The figure shows the adjusting of a valve yoke or bridge. The valve yoke is actuated by the rocker and valve lash is defined at the rocker adjusting screw. The adjustment shown ensures that both valves are actuated at exactly the same moment.

Answer B is incorrect. Valve lash is not adjusted with this screw.

Answer C is incorrect. The injectors are not shown in this figure.

Answer D is incorrect. The injector hold-down is not shown in this figure.

Question #111
Answer A is correct. Only Technician A is correct. A micrometer is commonly used to measure crankshaft journals.

Answer B is incorrect. Most diesel engine manufacturers would disapprove of using a wire wheel to clean up crankshaft journals.

Answer C is incorrect. Only Technician A is correct.

Answer D is incorrect. Only Technician A is correct.

Question #112
Answer C is correct. Both Technicians A and B are correct. Starter ring gears should be checked for damaged teeth, and replaced if damage is present. Vehicles with automatic transmissions use flexplate assemblies that bolt to both the torque converter and the crankshaft directly.

Answer A is incorrect. Technician B is also correct.

Answer B is incorrect. Technician A is also correct.

Answer D is incorrect. Both technicians are correct.

Question #113
Answer C is correct. Both Technicians A and B are correct. EG and PG should never be mixed because they are chemically incompatible and once mixing has taken place, it becomes impossible to measure the extent of antifreeze protection provided by the coolant. The mixture in itself will not cause any engine or cooling system problems, but it will be impossible to determine the antifreeze mixture strength with either a refractometer or a hydrometer. If a mixture of EG and PG is known to have taken place and the coolant, for whatever reasons, cannot be immediately replaced, use a refractometer with an EG and a PG scale and average the two readings.

Answer A is incorrect. Technician B is also correct.

Answer B is incorrect. Technician A is also correct.

Answer D is incorrect. Both technicians are correct.

Question #114
Answer B is correct. In this sequence, the technician should inspect the cylinder block bores first after a set of dry liners is removed. Sleeves should be removed with a puller and adapter plate or shoe. The procedure is obviously more simple on wet liners. Dry sleeves often require the use of mechanical, hydraulic, or air-over-hydraulic pullers. When these fail to extract a seized liner, vertical arc weld runs may break a liner free, but great care has to be exercised because the block may become distorted. To use this method of removing a seized liner, first ensure that every critical engine component is moved well out of the way, then use two vertical-down runs with an E6010/11 electrode at 90° from the piston thrust faces. Shock cool the welds with cold water and attempt to use the puller again. Avoid fracturing seized liners out of their bores as this practice almost always results in bore damage. As a last resort, liners can be machined out using a boring jig.

Answer A is incorrect. Main bearing caps can be left in place.

Answer C is incorrect. Oil ports do not need to be inspected immediately after liner removal.

Answer D is incorrect. Cam bearing bores are not necessarily inspected immediately after liner removal.

Question #115
Answer B is correct. Only Technician B is correct. If excessive contaminants are found in the fuel filters, the source of the contamination needs to be identified and corrected. If the source is the vehicle fuel tank, the tank should be drained, and flushed of the contamination.

Answer A is incorrect. The source of the contamination should be corrected before the vehicle is returned to service; otherwise the filters will only clog again.

C is incorrect. Only Technician B is correct.

D is incorrect. Only Technician B is correct.

Question #116
Answer C is correct. Both Technicians A and B are correct. Electronic connectors should never be probed for any reason. Breakout boxes and Ts are commonly used to test electronic circuits in modern vehicles.

Answer A is incorrect. Technician B is also correct.

Answer B is incorrect. Technician A is also correct.

Answer D is incorrect. Both technicians are correct.

Question #117
Answer A is correct. The bolt holes should be checked first; the remaining procedures are performed after the cylinder head has been installed. On in-line multicylinder heads, it is usually required that separate heads be aligned with a straightedge across the intake manifold faces before torquing. Failure to observe torquing increments and sequencing can result in cracked cylinder heads, failed head gaskets, and fire rings that will not seal. Cylinder head bolts should be installed lightly oiled. Excessive quantities of oil should be avoided because the excess can drain into the bolt hole and cause a hydraulic lock.

Answer B is incorrect. Head bolts are torqued after installation.

Answer C is incorrect. Rocker pedestals are checked after head installation.

Answer D is incorrect. Valve bridges are adjusted after installation.

Question #118
Answer A is correct. Only Technician A is correct. When a vibration occurrence rate seems to be half the engine speed, check the camshaft and cam trains first because the camshaft is rotated at half engine speed. Technician A is correct in suspecting a bent camshaft.

Answer B is incorrect. A bent connecting rod would produce a vibration incidence that would correlate with engine speed.

Answer C is incorrect. Only Technician A is correct.

Answer D is incorrect. Only Technician A is correct.

Question #119
Answer A is correct. The figure shows a hydraulically actuated, electronic unit injector. These injectors are actuated by high-pressure engine oil.

Answer B is incorrect. Electronics simply open or close an oil passage.

Answer C is incorrect. Electromechanical would be a EUI-style injector.

Answer D is incorrect. A mechanical injector would have no electronics associated with it and would not be pressurized by oil.

Question #120
Answer B is correct. Only Technician B is correct. When a code is produced so is a fault mode indicator (FMI), which can be read by diagnostic instruments capable of reading SAE codes.

Answer A is incorrect. Technician A is incorrect. Active codes can never be erased because the fault is currently occurring.

Answer C is incorrect. Only Technician B is correct.

Answer D is incorrect. Only Technician B is correct.

Question #121
Answer C is correct. Both Technicians A and B are correct. When turbocharger lines and hoses become nicked or frayed, it is good practice to replace them rather than repair them, so Technician A is correct. Technician B is also correct because it is good practice to inspect oil tubes for damage before returning them to service.

Answer A is incorrect. Technician B is also correct.

Answer B is incorrect. Technician A is also correct.

Answer D is incorrect. Both technicians are correct.

Question #122

Answer A is correct. Only Technician A is correct. Overfueling causes black smoke emission, essentially soot caused by partially combusted fuel.

Answer B is incorrect. Coolant in the combustion chamber condenses in the exhaust and is emitted as white smoke.

Answer C is incorrect. Only Technician A is correct.

Answer D is incorrect. Only Technician A is correct.

Question #123

Answer A is correct. The dash is most likely faulty.

Answer B is incorrect. Since the signal to the dash is feed off the data bus, the alternator is not the Most-Likely cause of the problem.

Answer C is incorrect. A faulty oil pressure sensor would effect only the oil pressure gauge, and it should also set a DTC in the engine ECM.

Answer D is incorrect. A faulty oil pressure gauge would only effect the oil pressure gauge, not the whole dash.

Question #124

Answer A is correct. 125°F is the cut-off point for electrolyte temperature during charging. Exceeding this temperature can result in battery damage or explosion.

Answer B is incorrect. That is above 125°F

Answer C is incorrect. That is below 125°F

Answer D is incorrect. That is above 125°F

Question #125

Answer D is correct. Neither technician is correct.

Answer A is incorrect. Technician A is incorrect. If connectors are damaged in any way, they should be replaced. Securing with a wire-tie is poor practice.

Answer B is incorrect. Technician B is incorrect. While the damaged connector should be replaced, the terminal pins should be inspected to determine if replacement is necessary. Terminal pin replacement is not necessary to replace the connector in most cases.

Answer C is incorrect. Neither technician is correct.

Question #126

Answer A is correct. Most manufacturers recommend that the snap ring gap be in the down position. Most important, it is never located at 90° to the direction of travel. All full-floating piston pins require a means of preventing the pin from exiting the pin boss and contacting the cylinder walls. Snap rings and plugs are used. When installing the internal snap rings used by most engine OEMs, observe the installation instructions.

Answer B is incorrect. The ring gap should not go up.

Answer C is incorrect. The ring gap should not go horizontal; this will allow inertia forces the opportunity to loosen the ring.

Answer D is incorrect. It does matter where the ring gap is installed.

Question #127

Answer C is correct. For the regulator to maintain a desired fuel pressure limit, the valve must be able to move in its bore, bleeding off excess pressure. If the valve was seized in the bore, the pressure can build to higher than normal pressure.

Answer A is incorrect. A broken valve spring would cause lower than normal fuel pressures.

Answer B is incorrect. A fuel tank vent controls air pressure in the fuel tank, not pump pressure.

Answer d is incorrect. Aerated fuel usually causes low fuel pressures.

Question #128

Answer A is correct. Only Technician A is correct. Since the "Wait To Start" lamp is illuminating, the ECM is commanding the heater grids on. The heater grids themselves may be suspect, so they should be tested against manufacturer's specifications. Generally, the grids will be checked for a specific resistance value. Technician A is correct.

Answer B is incorrect. The ECM controls the heater grid(s) with relay(s), but the relays should be tested for proper operation before they are replaced.

Answer C is incorrect. Only Technician A is correct.

Answer D is incorrect. Only Technician A is correct.

Question #129

Answer D is correct. Improper muffler, carbon build-up in the exhaust system, and a collapsed exhaust pipe are all capable of causing high exhaust back pressures. High exhaust temperature in itself would unlikely cause increased back pressure values so this is the LEAST-Likely cause.

Answer A is incorrect. An incorrect muffler can cause high EBP.

Answer B is incorrect. Carbon build-up can cause high EBP.

Answer C is incorrect. A collapsed exhaust pipe can cause high EBP.

Question #130

Answer C is correct. Both Technicians A and B are correct. Both technicians are describing the correct bench-test procedure for testing a cylinder head. A cylinder head should be heated to swell any cracks by flowing hot water through it for 10 minutes.

Answer A is incorrect. Technician B is also correct.

Answer B is incorrect. Technician A is also correct.

Answer D is incorrect. Both technicians are correct.

Question #131

Answer A is correct. Only Technician A is correct. Any visibly damaged connector should be replaced. Checking an electrical connector with an ohmmeter is not a good indicator of how the connector will perform when the circuit is energized. Every technician should understand the importance of voltage drop testing and that means understanding the limitations of the ohmmeter. If a copper starter motor cable has deteriorated to the extent that only 10 percent of its wire strands remain intact, an ohmmeter measurement will indicate that it checks out with a resistance reading similar to that of a cable in perfect condition because the ohmmeter is forcing a minute current through the test circuit. However, if a voltage drop test were to be performed on the cable while cranking the engine, the voltage dropped would immediately indicate the problem.

Answer B is incorrect. Placing electrical tape around the damaged connector is bad practice.

Answer C is incorrect. Only Technician A is correct.

Answer D is incorrect. Only Technician A is correct.

Question #132

Answer A is correct. Only Technician A is correct. Fan drive hub assemblies must be inspected. Clean the fan and related parts with clean solvent and dry them with compressed air; do not spin dry the bearings. Shielded bearings must not be washed as dirt may be washed in, and the cleaning fluid could not be entirely removed from the bearing. Examine the bearings for any indications of corrosion or pitting. Hold the inner race or cone so it does not turn, and revolve the outer race or cup slowly by hand. If rough spots are found, replace the bearings. Technician A is correct in describing how the fan bearings should be cleaned.

Answer B is incorrect. The fan bearings are not water-cooled.

Answer C is incorrect. Only Technician A is correct.

Answer D is incorrect. Only Technician A is correct.

Question #133

Answer D is correct. The LEAST-Likely cause of a failure to start an engine would be a low lubricant level. Low lube levels can result in significant engine damage. Although low oil levels can result in low oil pressure and engine shutdown, oil pressure and level-sensor inputs are ignored until after the engine starts.

Answer A is incorrect. Excessive exhaust back pressure could cause an engine to fail to start.

Answer B is incorrect. A short in the starter armature would cause slow cranking speed, which could cause the engine to fail to start. Normal minimum cranking speed is 100–200 rpm.

Answer C is incorrect. Battery terminal corrosion can cause slow cranking speed and a failure of the engine to start.

Question #134

Answer C is correct. Both Technicians A and B are correct. Engine overspeed can cause a crankshaft failure as can out-of-round main bearing bores.

Answer A is incorrect. Technician B is also correct.

Answer B is incorrect. Technician A is also correct.

Answer D is incorrect. Both technicians are correct.

Question #135

Answer B is correct. Only Technician B is correct. The O-rings used to seal wet liners are made from a variety of rubber-type compounds. It is important that OEM installation recommendations be observed. They may be installed dry, coated in coolant, soap, engine oil, and various other substances. Wet liners are often alloyed so they possess characteristics metallurgically superior to the block casting, which increase service life.

Answer A is incorrect. Wet sleeves or liners come into direct contact with coolant in the water jacket so they must seal top and bottom.

Answer C is incorrect. Only Technician B is correct.

Answer D is incorrect. Only Technician B is correct.

Question #136

Answer C is correct. Both Technicians A and B are correct. Excessive crankcase pressure can form an oily film on the inside of the engine door. If a turbocharger oil seal failure is present, oil can sometimes be seen on the exhaust joints.

Answer A is incorrect. Technician B is also correct.

Answer B is incorrect. Technician A is also correct.

Answer D is incorrect. Both technicians are correct.

Question #137

Answer B is correct. When removing the internal shutdown solenoid from a distributor type pump, the pump cover has to be removed to access the solenoid.

Answer A is incorrect. This style of shutdown solenoid does not have a solenoid cover.

Answer C is incorrect. The injector pump normally does not need to be removed for this repair.

Answer D is incorrect. The control module would not need to be removed for this repair.

Question #138

Answer D is correct. The cam plug seals the rear of the camshaft so this is not part of the valvetrain.

Answer A is incorrect. The rocker is part of the valvetrain.

Answer B is incorrect. The push tube (rod) is part of the valvetrain.

Answer C is incorrect. The cam follower (lifter) is part of the valvetrain.

Question #139

Answer C is correct. Both Technicians A and B are correct. Whenever injector sleeves or tubes are replaced, it is essential to check the installed injector tip protrusion to verify whether the sleeve has been properly installed in the cylinder head. Because the injector sleeve/tube comes into direct contact with coolant it is also good practice to perform a cylinder head, pressure leakage test (hydrostatic) to ensure there are no leaks.

Answer A is incorrect. Technician B is also correct.

Answer B is incorrect. Technician A is also correct.

Answer D is incorrect. Both technicians are correct.

Question #140

Answer D is correct. While a missing air filter can result in serious damage, this would have little limiting effect on power produced until the damage was caused, so this is the LEAST-Likely cause.

Answer A is incorrect. A restricted air filter would cause low pressure

Answer B is incorrect. A leaking fuel line would cause low power.

Answer C is incorrect. A restricted fuel filter would limit the amount of fuel the engine gets, resulting in low power.

Question #141

Answer C is correct. Both Technicians A and B are correct. Crosshead guide pins must be perfectly set at right angles to prevent side loading of the valves by the bridge, so Technician A is correct. Technician B is also correct in saying that crosshead guide pins should be measured to specification with a micrometer.

Answer A is incorrect. Technician B is also correct.

Answer B is incorrect. Technician A is also correct.

Answer D is incorrect. Both technicians are correct.

Question #142

Answer A is correct. Only Technician A is correct. If a thermostat is stuck in the open position, it can cause an engine to fail to reach the normal operating temperature window.

Answer B is incorrect. An overfilled radiator tank is unlikely to prevent an engine from reaching its normal operating temperature. It would most likely result in coolant being sent to the recovery bottle.

Answer C is incorrect. Only Technician A is correct.

Answer D is incorrect. Only Technician A is correct.

Question #143

Answer C is correct. Both Technicians A and B are correct. The alignment specifications for pulley-running centerlines as described are correct. Pulley belts should also not contact the bottom of the pulley grooves.

• When two or more identical belts are used on the same pulley, all of the belts must be replaced at the same time.

• Make sure the distance between the pulley centers is as short as possible when you install the belts. Do not roll the belts over the pulley. Do not use a tool to pry the belts onto the pulley.

• Pulleys must not be out of alignment more than 1/16 in. (1.59 mm) for each 12 in. (30.5 mm) of distance between the pulley center.

• Belts must not touch the bottom of the pulley grooves, and they must not protrude more than 3/32 in. (2.38 mm) above the outside diameter of the pulley.

• When identical belts are installed on a pulley, the protrusion of the belts must not vary more than 1/16 in. (1.59 mm).

• Make sure that the belts do not touch or hit any part of the engine.

Answer A is incorrect. Technician B is also correct.

Answer B is incorrect. Technician A is also correct.

Answer D is incorrect. Both technicians are correct.

Question #144

Answer C is correct. The throttle position sensor, a potentiometer, receives V-Ref, usually 5 VDC, and returns a portion of that voltage as a signal depending on accelerator pedal angle. To verify that the signal produced has actually passed through the sensor (as opposed to an open or shorted signal), they are designed to produce as a signal a small voltage value at 0 pedal travel and always less than V-Ref at maximum pedal travel. Typically, this would be the values indicated in Answer C.

Answer A is incorrect. The very ends of the voltage range 0 volts or 5 volts would indicate an open or short.

Answer B is incorrect. Nearly all TPS signals start low at 0 throttle and work their way up at higher throttle.

Answer D is incorrect. Nearly all TPS signals start low at 0 throttle and work their way up at higher throttle.

Question #145

Answer D is correct. U-joint failure is usually caused by driveshaft problems, and would be unaffected by the flywheel housing alignment.

Answer A is incorrect. Rear oil seal failure can be caused by improper flywheel alignment.

Answer B is incorrect. Flywheel bolt breakage can be caused by improper flywheel alignment.

Answer C is incorrect. Transmission damage can be caused by improper flywheel alignment.

Question #146

Answer B is correct. When troubleshooting electronic input circuit components, it is essential to consult the manufacturer's service literature.

Answer A is incorrect. V-ref voltage check is not the first step.

Answer C is incorrect. Simulating V-ref voltage would be much further down the troubleshooting tree; it would not be the first step.

Answer D is incorrect. While checking resistance of the sensor would be an early troubleshooting action, it would not be the first step.

Question #147

Answer A is correct. Only Technician A is correct. Most gauges use the position of the sending unit to vary the ground to the gauge, with a full tank providing very little ground to the gauge, while an empty tank providing a good ground. If the circuit to the sending unit or the resistor has an open, the gauge will read full by default.

Answer B is incorrect. If the sending unit float filled with fuel, it would sink to the bottom of the tank, resulting in a gauge reading of empty all the time.

Answer C is incorrect. Only Technician A is correct.

Answer D is incorrect. Only Technician A is correct.

Question #148

Answer C is correct. Knowledge of a procedure is being tested in this question. You can bet that disconnecting the ground cable is going to be one of the first steps in any procedure involving the replacement of an electrical component, because this renders the chassis electrically neutral.

Answer A is incorrect. The starter bolts would likely be the last step in removal.

Answer B is incorrect. The solenoid is integral to the starter and does not need to be removed.

Answer D is incorrect. The magnetic switch does not need to be removed with the starter.

Question #149

Answer A is correct. Generally, when the fuel injector timing is set, the injector should be in a static position when the cylinder is not on a compression stroke. If the position of the exhaust valves is open, the camshaft is 180° from firing the injector, and timing height can be set. Manufacturer's specifications and procedures should always be consulted first.

Answer B is incorrect. If the follower is fully depressed, the injector is firing, and cannot be properly adjusted.

Answer C is incorrect. If the cylinder is at TDC of a compression stroke, the injector is firing and cannot be properly adjusted.

Answer D is incorrect. Excessive valvetrain noise is usually an indication of excessive valve clearance, not improper fuel injector timing.

Question #150

Answer A is correct. Whenever a diesel engine emits black smoke, oxygen starvation caused by air-cleaner restriction should be checked first, mainly because it can be checked and quickly eliminated as a cause. Test the intake air inlet restriction using a water manometer. Specifications should always be checked to the OEM values, but typical maximum values will be close to:

- 15 inches H_2O vacuum—naturally aspirated engines
- 25 inches H_2O vacuum—boosted engines

Answer B is incorrect. The fuel pump (transfer pump) is very unlikely to cause a black-smoke condition.

Answer C is incorrect. The injector pump may cause black smoke, but it is not the first item to check.

Answer D is incorrect. The turbocharger can cause black smoke, but it is not the first item to check.

Question #151

Answer C is correct. A leaking intake manifold gasket is the LEAST-Likely cause because it would not cause fuel to enter the lubrication oil. When fuel contaminates engine oil, the oil loses its lubricity and it appears thinner and blacker in color. The condition is usually easy to detect because small amounts of fuel in oil can be recognized by odor. When fuel is found in significant quantities in engine oil, the cylinder head(s) are the likely source.

Answer A is incorrect. A cracked fuel gallery in a cylinder head could cause fuel to enter the lubricating oil.

Answer B is incorrect. A broken piston ring could cause incomplete combustion, which could cause fuel to enter the lubricating oil.

Answer D is incorrect. A leaking injector seal can cause fuel to enter the lubricating oil.

Question #152

Answer C is correct. Before removing each piston and connecting rod assemblies, scrape carbon deposits (ring ridge) from the upper-inside wall of each cylinder liner using a ridge reamer.

Answer A is incorrect. The crankshaft is not removed prior to the piston.

Answer B is incorrect. The piston pin cannot be removed prior to the piston assembly.

Answer D is incorrect. The cylinder sleeves are removed after the piston assembly.

Question #153

Answer C is correct. Both technicians are correct. Electronic unit injectors (EUIs) are mechanically actuated by an injector train and cam profile and metered fuel quantity is controlled by computer switching of the EUI solenoid.

Answer A is incorrect. Technician B is also correct.

Answer B is incorrect. Technician A is also correct.

Answer D is incorrect. Both technicians are correct.

Question #154

Answer D is correct. A coil spring, radiator temperature, and air temperature can all be used to regulate fan operation in a thermostatically controlled fan. Air pressure is not a factor in controlling fan cycling so this is the exception. Thermatic viscous drive fan hubs are integral fan drive units with no external controls. They use silicone fluid as a drive medium between the drive hub and the fan drive plate.

Answer A is incorrect. This fan does use a coil spring.

Answer B is incorrect. Air temperature does regulate this fan operation.

Answer C is incorrect. Radiator temperature does affect this fan because radiator temperature affects the temperature of the air passing through it.

Question #155

Answer B is correct. Only Technician B is correct. Technician B has identified the cause of the driver's complaint: progressive shifting has been programmed into customer data programming to improve fuel economy. Progressive shifting can assist less experienced drivers of buses by providing engine speed prompts during different vehicle speeds. Progressive shifting can make a significant difference to vehicle fuel economy.

Answer A is incorrect. Technician A is incorrect. Fuel map programming is factory set and can only be changed when new calibration files are downloaded to the engine ECM.

Answer C is incorrect. Only Technician B is correct.

Answer D is incorrect. Only Technician B is correct.

Question #156

Answer A is correct. Timing an EUI requires the setting of the EUI plunger height, a mechanical adjustment performed by setting the rocker arm adjusting screw.

Answer B is incorrect. Calibration codes do not set timing.

Answer C is incorrect. Timing codes are not programmed.

Answer D is incorrect. The cam follower is not adjusted.

Question #157

Answer B is correct. Only Technician B is correct. Excessive use of ether may result in broken piston rings and damaged piston ring lands. The automatic ether injection system is an optional starting aid. It is designed to spray a controlled flow of starting fluid into the air intake system during engine cranking in cold weather. Do not keep the starter engaged for more than 15 seconds at a time, and wait at least 30 seconds before trying again. With the automatic ether injection system, the engine should start in approximately 6 seconds, so cranking time should be kept short. If the engine will not start after several short cold cranking cycles, the ether delivery system and its components should be checked. If the problem is not with the ether delivery system, the "no start" condition should be diagnosed to avoid overinjecting the engine with ether.

Answer A is incorrect. Ether should never be used with glow plugs. Vehicles equipped with glow plugs have warning labels warning against severe damage if ether is used.

Answer C is incorrect. Only Technician B is correct.

Answer D is incorrect. Only Technician B is correct.

Question #158

Answer B is correct. Only Technician B is correct. It is not uncommon practice to recore a turbocharger by replacing the center housing which is a prebalanced assembly. Recoring can only be performed if both the turbine and impeller housings are in good condition.

Answer A is incorrect. Turbochargers are disassembled in the field.

Answer C is incorrect. Only Technician B is correct.

Answer D is incorrect. Only Technician B is correct.

Question #159

Answer C is correct. Both Technicians A and B are correct. Technician A's method of cleaning a wet liner is a good one. Technician B is also correct in describing how to check for a cylinder out-of-round condition.

Answer A is incorrect. Technician B is also correct.

Answer B is incorrect. Technician A is also correct.

Answer D is incorrect. Both technicians are correct.

Question #160

Answer B is correct. Oil coolers are heat exchangers consisting of a housing and a bundle (element/core) through which oil is pumped and around which coolant is circulated. Oil temperatures in diesel engines run higher than coolant temperatures, typically around 110°C (230°F). However, the engine coolant reaches its operating temperature more rapidly than the oil and plays a role in heating the oil to operating temperature in cold weather startup/warmup conditions. The figure shows an oil cooler bundle viewed from the header end.

Answer A is incorrect. This is not an oil filter.

Answer C is incorrect. This is not filter housing.

Answer D is incorrect. This is not an oil pump.

Question #161

Answer A is correct. Only Technician A is correct. Valve rotators are used on diesel engine valves primarily to prevent carbon build-up on the valve seats.

Answer B is incorrect. Valve rotators do not in any way help seal the valve guide.

Answer C is incorrect. Only Technician A is correct.

Answer D is incorrect. Only Technician A is correct.

Question #162

Answer D is correct. Neither technician is correct.

Answer A is incorrect. Technician A is incorrect. At idle speeds, the turbocharger is spinning at its slowest, and bearing load is light. As long as the turbocharger is receiving adequate lubrication, the amount of idle time should not be causing bearing damage.

Answer B is incorrect. Technician B is incorrect. While a blocked oil supply line can cause oil starvation and bearing failure, the oil line to the turbocharger should be inspected to determine if replacement is necessary. Replacing parts without checking for failure first is poor practice.

Answer C is incorrect. Neither technician is correct.

Question #163

Answer A is correct. Only Technician A is correct. A starter solenoid has two sets of windings or coils, the pull-in and hold-in windings.

Answer B is incorrect. The pull-in winding draws more current than the hold-in winding, simply because it takes more force to pull the starter pinion into the engaged position than it does to hold it in the engaged position.

Answer C is incorrect. Only Technician A is correct.

Answer D is incorrect. Only Technician A is correct.

Question #164

Answer C is correct. Both Technicians A and B are correct. Technician A is correct because antifreeze does not readily evaporate, making it somewhat easy to detect small leaks. Technician B is also correct because repeated engine overheating can shorten the life of a thermostat. Cooling system leakage is common, and the system should be inspected by the operator daily. Cold leaks may be caused by contraction of mated components at joints, especially hose clamps. Cold leaks often cease to leak at operating temperatures. Many fleet operators replace all the coolant after a prescribed in-service period regardless of its appearance to avoid the costs incurred in a breakdown. Silicone hoses are more expensive than the rubber compound type but they usually have longer service life. Silicone hoses require the use of special clamps, and these are sensitive to overtightening. They must be torqued to the required specification. Pressure testing a cooling system will locate most external cooling system leaks. A typical cooling system pressure testing kit consists of a hand-actuated pump and gauge assembly calibrated to 625 psi (170 kPa) plus various adapters for the different types of fill necks and radiator caps. Some are capable of vacuum testing.

Answer A is incorrect. Technician B is also correct.

Answer B is incorrect. Technician A is also correct.

Answer D is incorrect. Both technicians are correct.

Question #165

Answer A is correct. Only Technician A is correct. Depending on the location of the coolant level sensor, the coolant level may drop below sensor, triggering the fault. Once the coolant in the reservoir levels out, the fault will become inactive.

Answer B is incorrect. Low coolant level alerts are time lagged in that no displayed warning is produced by the ECM until a specific time period has lapsed: 8 seconds would be typical. Currently, most electronically managed engines have radiators equipped with low coolant level warning systems. Most operate using the same principles. The ECM outputs a signal to a probe (or sensor), which grounds through the coolant. The probe is usually located in the top radiator tank. When the probe fails to ground through the coolant, a low coolant level warning is generated. The outcome depends on how the ECM has been programmed (this is a customer data program option). In most cases, a lag (from 5 to 12 seconds) is required before the ECM resorts to a programmed failure strategy, which may be to alert the operator, ramp down to a default rpm/engine load, or shut down the engine after a suitable warning period.

Answer C is incorrect. Only Technician A is correct.

Answer D is incorrect. Only Technician A is correct.

Question #166

Answer C is correct. Low voltage, a short in the wiring to the transducer, and a defective transducer will all cause inaccurate readings in an electric oil pressure gauge. However, low oil pressure in itself would not cause an inaccurate reading at the gauge. Of all the engine monitoring devices used on an engine, the oil pressure sensor is one of the most critical. Loss of engine oil pressure in most cases will cause a nearly immediate engine failure. Years ago when few of an engine's operating conditions were monitored and displayed to the operator, there was always a means of signaling a loss of oil pressure. Several types of sensors are used in today's engines, including: variable capacitance (pressure) sensor, piezoelectric, bourdon gauge, and an electrical gauge. Whichever is used to display oil pressure, none should be used as the only means of diagnosing low oil pressure. Use a good quality master gauge such as a gauge with a fluid-filled display dial. CAUTION: When disconnecting high-pressure oil lines on HEUI high-pressure oil pumps, relieve the system pressure first; wear safety glasses when performing this procedure to avoid personal injury.

Answer A is incorrect. Low voltage would cause the gauge to read inaccurate.

Answer B is incorrect. A short in the transducer would cause the gauge to read incorrectly.

Answer D is incorrect. A defective transducer would cause the gauge to read incorrectly.

Question #167

Answer C is correct. Both Technicians A and B are correct. Cylinder hydraulic lock is usually caused by liquid (generally engine coolant) filling the cylinder above the piston, preventing the engine from rotating. Both the intercooler (such as would be used on a DDC 2-stroke cycle engine) and a crack in a cylinder head are possible sources of coolant that fills a cylinder.

Answer A is incorrect. Technician B is also correct.

Answer B is incorrect. Technician A is also correct.

Answer D is incorrect. Both technicians are correct.

Question #168

Answer C is correct. The first step is to check the belts for cracking and wear. If worn or there is evidence of cracks, the belts should be changed, not adjusted. Fan pulleys use external V or poly-V grooves and internal bearings of the roller, taper roller, and bushing types. The belt tension should be adjusted using a belt tensioner.

Answer A is incorrect. Drive pulleys are not replaced just because the belts are being adjusted.

Answer B is incorrect. Idler pulleys are not replaced just because the belts are being adjusted.

Answer D is incorrect. Belt dressing compound is not typically approved by OEMs.

Question #169

Answer C is correct. Both Technicians A and B are correct. Thermostats allow the engine to reach and maintain operating temperature. If thermostats were not installed, the coolant would continuously circulate through the radiator and never reach proper operating temperature. Diesel engines tend to waste fuel at lower temperatures due to incomplete combustion. Operating the engine below normal temperature will cause increased fuel consumption. If the fuel line were leaking, fuel would be wasted and cause fuel mileage to decrease. Both technicians are correct.

Answer A is incorrect. Technician B is also correct.

Answer B is incorrect. Technician A is also correct.

Answer D is incorrect. Both technicians are correct.

Question #170

Answer D is correct. Replacing the bushings is part of starter motor overhaul, not part of installing it.

Answer A is incorrect. Control wiring would be reinstalled.

Answer B is incorrect. The motor would be aligned.

Answer C is incorrect. The mounting bolts must be reinstalled.

Question #171

Answer A is correct. Only Technician A correctly describes the recommended practice process of selectively fitting a set of dry liners to a cylinder block.

Answer B is incorrect. Crosshatch is not required to be cut to the cylinder block bore in any engine that uses liners; this is machined to the liner bore. The dry liner uses thinner-walled sleeves than the wet liner. Dry liners (sleeves) are installed into the block bore, usually with a marginally loose fit, and are retained by the cylinder head. The dry sleeves do not transfer heat as efficiently as the wet liners, but they are easily replaced and do not present coolant-sealing problems.

Answer C is incorrect. Only Technician A is correct.

Answer D is incorrect. Only Technician A is correct.

Question #172

Answer D is correct. The figure shows a gear-type engine oil pump and pressure relief valve assembly. Gerotor-type oil pumps use an internal crescent gear pumping principle. An internal impeller with external crescent vanes is rotated within an internal crescent gear, also known as a rotor ring. The inner rotor or impeller has one less lobe than the rotor ring. As the inner rotor is driven within the outer rotor, only one lobe is engaged at any given moment of operation. In this way, oil from the inlet port is picked up in the crescent formed between two lobes on the impeller. As the impeller rotates, oil is forced through the outlet port as the lead lobe once again engages. The assembly is rotated within the Gerotor pump body.

Answer A is incorrect. This is not a fuel transfer pump.

Answer B is incorrect. This is not the water pump.

Answer C is incorrect. This is not the fuel injection pump.

Question #173

Answer C is correct. Diesel exhaust normally contains small amounts of soot and ash. Black sooty deposits in and around the engine compartment are telltale signs of exhaust system leaks.

A is incorrect. Excessive idling will not cause soot build-up inside the engine compartment.

B is incorrect. Although a malfunctioning boost pressure sensor can cause overfueling and excess soot, the soot would normally exit through the tailpipe and not collect in the engine compartment.

Answer D is incorrect. Loose air intake hose clamps would allow dirt to enter the intake air stream, not produce soot deposits.

Question #174

Answer D is correct. Suction circuit leaks, improperly set throttle lever angle, and clogged fuel filters can all result in a low power condition. However, a leak in the injector return circuit is unlikely to cause low power, so it is the LEAST-Likely cause of the problem.

Answer A is incorrect. Suction circuit leaks cause low power.

Answer B is incorrect. A defective fuel transfer pump will cause low power.

Answer C is incorrect. A clogged fuel filter will cause low power.

Question #175

Answer B is correct. Only Technician is B correct. Indications of corrosion in any sealing surface of a wet liner must be addressed or the liner will leak when installed.

Answer A is incorrect. If the rust and scale are not removed, the liner O-rings will most likely leak.

Answer C is incorrect. Only Technician B is correct.

Answer D is incorrect. Only Technician B is correct.

Question #176

Answer C is correct. The MUI rack controls the plunger effective stroke so it will not cause a spray pattern problem.

Answer A is incorrect. Nozzle-tip holes can cause an improper spray pattern.

Answer B is incorrect. The injection nozzle valve can cause improper spray patterns.

Answer D is incorrect. A worn injector follower can cause no spray from the injector.

Question #177

Answer B is correct. Voltage drops test resistance in an energized circuit, and as such they are the most meaningful method of checking active circuits. Only a voltmeter can be used to perform this test. Getting used to using your DMM on real-world circuits means testing properly functioning circuits. Too often, a technician tests a circuit for a defect without knowing how the circuit would test if it was fully functional. Make a practice of testing good circuits as well as bad. Remember, your DMM is capable of making exact measurements and the specific value is significant. When evaluating circuit performance, there is a difference between 12.6 and 12.5 volts; make a habit of writing down the results of circuit testing.

Answer A is incorrect. A shunt resistor would not be used.

Answer C is incorrect. An ohmmeter cannot measure voltage.

Answer D is incorrect. A millimeter is a distance measurement.

Question #178

Answer A is correct. Using a stethoscope, shorting out hydromechanical injectors, and performing an electronic cylinder cut-out test are all valid procedures when attempting to troubleshoot an engine miss. However, attempting to locate an engine miss by listening to the exhaust is unlikely to locate which cylinder is misfiring so this is the LEAST-Likely to produce satisfactory results.

Answer B is incorrect. Using a stethoscope is a valid method of identifying an engine miss.

Answer C is incorrect. Manual shorting out (depressing) a hydromechanical injector is a good test method.

Answer D is incorrect. An electronic cut-out test is a valid test method.

Question #179

Answer B is correct. Using a chisel to remove valve seats is not good practice because the cylinder head can be damaged; the rest of the answers reference appropriate steps when performing cylinder valve service. Most current truck diesel engines use valve seat inserts rather than integral valve seats machined into the cylinder head. Valve seat inserts are press-fit to a machined recess in the cylinder head.

Since most of the heat of the valve must be transferred from the valve to the seat, it is essential that the contact area of the seat and the cylinder head be maximized. A sectored collet that expands into the valve seat is used to remove the valve seat. The seat can either be levered out or driven out with a slide hammer. When installing new valve seats, the seat counterbore must be first cleaned out using a low-abrasive emery cloth.

Answer A is incorrect. A valve seat can be struck lightly with the ball-peen end of a hammer. If the seat "rings" it is loose.

Answer C is incorrect. Valve seats are checked for concentricity with the guide.

Answer D is incorrect. Valve grinding stones should be dressed prior to being used.

Question #180
Answer C is correct. Both Technicians A and B are correct. When pressure testing a cylinder block for suspected cracks, immersing it in hot water for 20–30 minutes should be sufficient to swell any cracks to observe a leak. Most diesel engine manufacturers recommend that a cracked cylinder block be replaced rather than repaired.

Answer A is incorrect. Technician B is also correct.

Answer B is incorrect. Technician A is also correct.

Answer D is incorrect. Both technicians are correct.

Question #181
Answer D is correct. The figure shows a valve stem stretched at the fillet.

Answer A is incorrect. The figure shows a valve stem stretched at the fillet.

Answer B is incorrect. The figure shows a valve stem stretched at the fillet.

Answer C is incorrect. The figure shows a valve stem stretched at the fillet.

Question #182
Answer D is correct. The tool shown is designed to cut a cylinder liner counterbore. Check the cylinder sleeve counterbore for the correct depth and circumference. Counterbore depth should typically not vary by more than 0.001 in. Counterbore depth can be subtracted from the sleeve flange dimension to calculate sleeve protrusion. Recut and shim the counterbore using OEM-recommended tools and specifications.

Answer A is incorrect. This is not a ridge remover.

Answer B is incorrect. This is not a liner remover.

Answer C is incorrect. This machine does not machine the crosshatch.

Question #183
Answer A is correct. Only Technician A is correct. Most exhaust brakes are actuated using compressed air supplied by the vehicle's air system. If system pressure to the actuator is insufficient to close the valve, the brake will not work. Technician A is correct in checking that the proper amount of air is available.

Answer B is incorrect. If the valve's return spring was broken, the brake would apply, but it would most likely not release.

Answer C is incorrect. Only Technician A is correct.

Answer D is incorrect. Only Technician A is correct.

Question #184
Answer B is correct. Only Technician B is correct. Advanced fuel injection timing damages the outer edges of the piston and melts, eroding the top ring land above the Ni-resist insert. The erosion is caused by fuel dispersal/condensation outside of the combustion bowl.

Answer A is incorrect. Fuel with low CN is likely to produce failures consistent with retarded timing.

Answer C is incorrect. Only Technician B is correct.

Answer D is incorrect. Only Technician B is correct.

Question #185

Answer C is correct. Both Technicians A and B are correct. When high-pressure fuel injection lines are removed from an engine, it is essential to immediately cap them to prevent the entry of dirt and dust. Technician A is correct. Low power in PLN fuel systems can be caused by overtorquing of high-pressure line nuts, which collapses the sealing nipple, reducing flow potential. So Technician B is also correct.

Answer A is incorrect. Technician B is also correct.

Answer B is incorrect. Technician A is also correct.

Answer D is incorrect. Both technicians are correct.

Question #186

Answer B is correct. Only Technician B is correct. Fuel to be injected is metered by the rail actuator solenoid. Fuel is directed into and out of the HPI-TP injectors by means of 3 exterior annuli separated by 4 O-ring seals. The upper annulus feeds timing fuel to the injector, the middle annulus discharges drain/spill fuel, and the lower annulus receives the metering fuel.

Answer A is incorrect. The diagram shows an HPI-TP injector. The fuel passing through the timing orifice acts as a hydraulic slug to define timing and none of it will be injected; it is routed back to the fuel tank.

Answer C is incorrect. Only Technician B is correct.

Answer D is incorrect. Only Technician B is correct

Question #187

Answer C is correct. Both Technicians A and B are correct. Technician A correctly describes how to remove the surface charge from a battery that has been recently charged. Also, a stored battery does not have a surface charge.

Answer A is incorrect. Technician B is also correct.

Answer B is incorrect. Technician A is also correct.

Answer D is incorrect. Both technicians are correct.

Question #188

Answer C is correct. Both technicians are correct. By-pass or centrifugal oil filters should be serviced at the same interval as the oil and full-flow filters are changed. This will ensure the centrifuge is operating at peak efficiency. A full bowl or sudden increase in the amount of material collected in the centrifuge bowl can indicate excessive contaminants entering the engine oil. This can be caused by cracked or worn piston rings, excessive blow-by, or poor air filtration.

Answer A is incorrect. Technician B is also correct.

Answer B is incorrect. Technician A is also correct.

Answer D is incorrect. Both technicians are correct.

Question #189

Answer D is correct. All of the checks are valid EXCEPT for the one referencing the interference fit of the liner into the cylinder bore. Wet sleeves are never interference fitted to the cylinder block bore; they are sealed with O-rings. Dry sleeves are interference fit.

Answer A is incorrect. The contact surfaces should be inspected.

Answer B is incorrect. The O-ring grooves should be inspected.

Answer C is incorrect. The bore should be inspected for scoring.

Question #190

Answer B is correct. Only Technician B is correct. The pump shown in the center of the V cradle in this figure is the HEUI high-pressure oil pump that pressurizes engine oil to actuate the HEUI injectors. If this pump is producing lower than specified pressure, one of the results would be low power.

Answer A is incorrect. The pump does not supply fuel to the injectors.

Answer C is incorrect. Only Technician B is correct.

Answer D is incorrect. Only Technician B is correct.

Question #191

Answer B is correct. Only Technician B is correct. Crankcase pressure should be tested across a range of rpms and engine loads usually under monitored test conditions such as would be found on a dynamometer. The highest crankcase pressures are produced when engine cylinder pressures are high and engine speeds are low (more time for cylinder blowby each cycle).

Answer A is incorrect. The engine is normally run at varying rpms.

Answer C is incorrect. Only Technician B is correct.

Answer D is incorrect. Only Technician B is correct.

Question #192

Answer A is correct. Only Technician A is correct. When boosting a vehicle using another, it is good practice to switch off the engine of the boost vehicle while making the battery connections to prevent electrical/electronic damage.

Answer B is incorrect. Most bus electronically managed engines are designed to sustain transient (short-term) voltage overloads such as would be caused by a portable generator, a common start assist tool.

Answer C is incorrect. Only Technician A is correct.

Answer D is incorrect. Only Technician A is correct.

Question #193

Answer C is correct. Both Technicians A and B are correct. Jacket water aftercoolers (JWAC) use a split design intake manifold that contains the aftercooler core. The water pump pumps engine coolant through the aftercooler core. Incoming air moving from the compressor side of the turbocharger through the intake manifold is at approximately 300°F (148°C) before going through the aftercooler core. This action reduces the air temperature to approximately 195°F (91°C), thereby increasing its density. Both air locked in the intercooler and clogged air tubes could cause lack of power and higher coolant temperatures.

Answer A is incorrect. Technician B is also correct.

Answer B is incorrect. Technician A is also correct.

Answer D is incorrect. Both technicians are correct.

Question #194

Answer C is correct. Technicians A and B are correct. When a crankshaft fails, especially if the indications are that it is a torsional failure, everything on its power train (driven by it) should be closely inspected. Both technicians are correct because the vibration damper and flywheel are among the first components that should be inspected. Torsional stresses are the twisting vibrations that a crank is subjected to at high speed. Crankshaft torsional vibration occurs because while under compression (that is, driving the piston assembly attached to it upward on the compression stroke) a given crank throw will slow to a speed marginally less than average crank speed. Subsequently, when the throw receives a power stroke, it will accelerate to a speed marginally greater than average crank speed. These twisting vibrations or oscillations take place at high frequencies, and crankshaft design, materials, and hardening methods must take them into account. Torsional stresses tend to peak at crank journal oil holes at the flywheel end of the shaft. Torsional oscillations are amplified when an engine is run at slower speeds with high cylinder pressures, because the real-time duration between cylinder firing pulses is extended. This type of running (lower speed/high load) is known as lugging, but with many new generation, high torque-rise engines, engine torsional oscillations can be projected through the drivetrain.

Answer A is incorrect. Technician B is also correct.

Answer B is incorrect. Technician A is also correct.

Answer D is incorrect. Both technicians are correct.

Question #195

Answer B is correct. Only Technician B is correct. The system has two warning lights. The check engine light signals a fault or intermittent fault to the driver. The stop engine light, which illuminates when the ECM has detected a problem that could result in serious engine failure, may be programmed to shut down the engine or ramp down to idle after the ECM detects problem.

Answer A is incorrect. A flashing SEL light would indicate that the idle shutdown timer had timed out.

Answer C is incorrect. Only Technician B is correct.

Answer D is incorrect. Only Technician B is correct.

Question #196

Answer B is correct. The cam bearing journal diameter is being measured.

Answer A is incorrect. The cam lobe is not being measured.

Answer C is incorrect. Cam lift is not being measured.

Answer D is incorrect. Cam duration is not being measured.

Question #197

Answer D is correct. Excessive use of ether to start an engine causes much higher cylinder pressures that can blow out piston ring lands.

Answer A is incorrect. Ether is unlikely to be the cause of an engine runaway because it ignites early and combusts very rapidly.

Answer B is incorrect. Ether will cause vapor lock, not hydraulic lock.

Answer C is incorrect. Ether will not damage the fuel system; ether is introduced into the air system.

Question #198

Answer B is correct. When the diagnostics determine that the ECM has to be changed, the customer data programming should be downloaded (and printed off if required) so this can be loaded onto the replacement ECM.

Answer A is incorrect. The battery should be disconnected during ECM replacement, however, the customer data should be downloaded first.

Answer C is incorrect. Injection lines do not need to be removed to replace the ECM.

Answer D is incorrect. The key should be off during ECM replacement.

Glossary

Actuator Device that delivers motion in response to an electrical signal.

A/D Converter Abbreviation for Analog-to-Digital Converter.

Additive An additive intended to improve a certain characteristic of the material or fluid.

Aftercooler Charge air cooling device, usually water cooled.

Air Compressor Engine-driven mechanism for supplying high-pressure air to the truck brake system.

Air Filter Device that minimizes the possibility of impurities entering the intake system.

Altitude Compensation System Altitude barometric switch and solenoid used to provide better drivability at 1000 feet plus above sea level.

Ambient Temperature Temperature of the surrounding air. Normally, it is considered to be the temperature in the service area where testing is taking place.

Amp Ampere.

Ampere Unit for measuring electrical current.

Analog Signal Voltage signal that varies within a given range (from high to low, including all points in between).

Analog-to-Digital Converter (A/D converter) Device that converts analog voltage signals to a digital format; located in the ECM.

Analog Volt/Ohmmeter (AVOM) Test meter used for checking voltage and resistance. Analog meters should not be used on solid state circuits.

Antifreeze Mixture added to water to lower its freezing point.

Armature Rotating component of a (1) starter or other motor. (2) Generator.

Articulation Pivoting movement.

ASE Automotive Service Excellence, a trademark of National Institute for Automotive Service Excellence.

Atmospheric Pressure Weight of the air at sea level; 14.696 pounds per square inch (psi) or 101.33 kilopascals (kPa).

Axis of Rotation Centerline around which a gear or part revolves.

Battery Terminal Tapered post or threaded studs on top of the battery case for connecting the cables.

Bimetallic Two dissimilar metals joined together that have different bending characteristics when subjected to changes of temperature.

Blade Fuse Type of fuse having two flat male lugs for insertion in female box connectors.

Blower Fan Fan that pushes or blows air through a ventilation, heater, or air conditioning system.

Boss Outer race of a bearing.

Bottoming Condition that occurs when the teeth of one gear touch the lowest point between teeth of a mating gear.

British Thermal Unit (Btu) Measure of heat quantity equal to the amount of heat required to raise 1 pound of water 1°F.

Btu British Thermal Unit.

CAA Clean Air Act.

Cartridge Fuse Type of fuse having a strip of low melting point metal enclosed in a glass tube. If an excessive current flows through the circuit, the fuse element melts at the narrow portion, opening the circuit and preventing damage.

Caster Angle formed between the kingpin axis and a vertical axis as viewed from the side of the vehicle. Caster is considered positive when the top of the kingpin axis is behind the vertical axis.

Cavitation Condition caused by bubble collapse.

C-EGR Cooled exhaust gas recirculation.

CFC Chlorofluorocarbon

Charging Circuit Alternator (or generator) and associated circuit used to keep the battery charged and power the vehicle electrical system when the engine is running.

Charging System System consisting of the battery, alternator, voltage regulator, associated wiring, and the electrical loads of a vehicle. The purpose of the system is to recharge the battery whenever necessary and to provide the current required to power the electrical components.

Check-Valve Valve that allows air to flow in one direction only.

Climbing Gear problem caused by excessive wear in gears, bearings, and shafts whereby the gears move sufficiently apart to cause the apex of the teeth on one gear to climb over the apex of another gear.

Clutch Device for connecting and disconnecting the engine from the transmission.

Coefficient of Friction Measurement of the amount of friction developed between two objects in physical contact when one of the objects is drawn across the other.

Coil Springs Spring steel spirals.

Compression Applying pressure to a spring or fluid.

Compressor Mechanical device that increases pressure within a circuit.

Condensation Process by which gas (or vapor) changes to a liquid.

Conductor Any material that permits the electrical current to flow.

Coolant Heater Component used to aid engine starting and reduce the wear caused by cold starting.

Coolant Hydrometer Tester designed to measure coolant specific gravity and determine antifreeze protection.

Coolant Liquid that circulates in an engine cooling system.

Cooling System System for circulating coolant.

Crankcase Housing within which the crankshaft rotates.

Cranking Circuit Starter circuit, including battery, relay (solenoid), ignition switch, neutral start switch (on vehicles with automatic transmission), and cables and wires.

Cycling (1) On-off action of the air conditioner compressor. (2) Repeated electrical cycling that can cause the positive plate material to break away from its grids and fall into the sediment base of the battery case.

Dampen To slow or reduce oscillations or movement.

Dampened Discs Discs that have dampening springs incorporated into the disc hub. When engine torque is transmitted to the disc, the plate rotates on the hub, compressing the springs. This action absorbs the torsional vibration caused by today's low rpm, high torque, engines.

Data Links Circuits through which computers communicate with other electronic devices such as control panels, modules, sensors, or other computers.

Deburring To remove sharp edges from a cut.

Deflection Bending or moving to a new position as the result of an external force.

DER Department of Environmental Resources.

Detergent Additive Additive that helps keep metal surfaces clean and prevents deposits. These additives suspend particles of carbon and oxidized oil in the oil.

Diagnostic Flow Chart Chart that provides a systematic approach to the electrical system and component troubleshooting and repair. These charts are found in service manuals and are vehicle-make and -model specific.

Dial Caliper Measuring instrument capable of taking inside, outside, depth, and step measurements.

Digital Binary Signal Signal that has only two values, on and off.

Digital Volt/Ohmmeter (DVOM) Test meter recommended for use on solid state circuits.

Diode Semiconductor Device formed by joining P-type semiconductor material with N-type semiconductor material. A diode allows current to flow in one direction, but not in the opposite direction.

DOT Department of Transportation.

Drive or Driving Gear Gear that drives another gear.

Driveline Propeller or driveshaft, and universal joints that link the transmission output to the axle pinion gear shaft.

Driveline Angle Alignment of the transmission output shaft, driveshaft, and rear axle pinion centerline.

Driven Gear Gear that is driven by a drive gear, by a shaft, or by some other device.

Driveshaft Assembly of one or two universal joints connected to a shaft or tube; used to transmit torque from the transmission to the differential.

Drivetrain Assembly that includes all torque transmitting components from the rear of the engine to the wheels.

ECM Electronic control module.

ECU Electronic control unit.

Eddy Current Circular current produced inside a metal core in the armature of a starter motor. Eddy currents produce heat and are reduced by using a laminated core.

EGR Exhaust gas recirculation. Method of rerouting exhaust gas into the combustion chamber to reduce NOx emissions.

Electricity Movement of electrons from one location to another.

Electromotive Force (EMF) Force that moves electrons between atoms. This force is the pressure that exists between the positive and negative points. It is measured in units called volts.

Electronically Erasable Programmable Memory (EEPROM) Computer memory that enables write-to functions.

Electrons Negatively charged particles orbiting every nucleus.

EMF Electromotive force.

Engine Brake Hydraulically operated device that converts the engine into a power-absorbing mechanism.

Environmental Protection Agency An agency of the U.S. government charged with the responsibilities of protecting the environment.

EPA Environmental Protection Agency.

EUI Electronic unit injector. A cam-actuated, electronically controlled fuel injector. Able to generate high-pressure required for fuel injection.

Exhaust Brake A slide mechanism which restricts the exhaust flow, causing exhaust back pressure to build up in the engine's cylinders. The exhaust brake actually transforms the engine into a power absorbing air compressor driven by the wheels.

False Brinelling Polishing of a surface that is not damaged.

Fatigue Failures Progressive destruction of a shaft or gear teeth usually caused by overloading.

Fault Code Code that is recorded into the computer's memory.

Federal Motor Vehicle Safety Standard (FMVSS) Federal standard that specifies that all vehicles in the United States be assigned a Vehicle Identification Number (VIN).

Fixed Value Resistor Electrical device designed to have only one resistance rating, which should not change, for controlling voltage.

Flammable Any material that will easily catch fire or explode.

Flare To spread gradually outward in a bell shape.

Foot-Pound English unit of measurement for torque. One foot-pound is the torque obtained by a force of 1 pound applied to a foot-long wrench handle.

Fretting Result of vibration that the bearing outer race can pick up the machining pattern.

Fuse Link Short length of smaller gauge wire installed in a conductor, usually close to the power source.

Fusible Link Term often used for an insulated fuse link.

Gear Disk-like wheel with external or internal teeth that serves to transmit or change motion.

Gear Pitch Number of teeth per given unit of pitch diameter, an important factor in gear design and operation.

Ground Negatively charged side of a circuit. A ground can be a wire, the negative side of the battery, or the vehicle chassis.

Grounded Circuit Shorted circuit that causes current to return to the battery before it has reached its intended destination.

Harness and Harness Connectors The vehicle's electrical system wiring providing a convenient starting point for tracking and testing circuits.

Hazardous Materials Any substance that is flammable, explosive, or is known to produce adverse health effects in people or the environment.

Heads-Up Display (HUD) Technology used in some vehicles that superimposes data on the driver's normal field of vision. The operator can view the information, which appears to "float" just above the hood at a range near the front of a conventional tractor or truck. This allows the driver to monitor conditions such as road speed without interrupting his view of traffic.

Heater Control Valve Valve that controls the flow of coolant into the heater core from the engine.

Heat Exchanger Device used to transfer heat, such as a radiator or condenser.

Heavy-Duty Truck Truck that has a GVW of 26,001 pounds or more.

High-Resistant Circuits Circuits that have an increase in circuit resistance, with a corresponding decrease in current.

High-Strength Steel Low-alloy steel that is stronger than hot-rolled or cold-rolled sheet steels.

Hinged Pawl Switch Simplest type of switch; one that makes or breaks the current of a single conductor.

HUD Heads-up display.

Hydrometer Tester designed to measure the specific gravity of a liquid.

Inboard Toward the centerline of the vehicle.

In-Line Fuse Fuse that is in series with the circuit in a small plastic fuse holder, not in the fuse box or panel. It is used, when necessary, as a protection device for a portion of the circuit even though the entire circuit may be protected by a fuse in the fuse box or panel.

Installation Templates Drawings supplied by some vehicle manufacturers to allow the technician to correctly install the accessory. The templates available can be used to check clearances or to ease installation.

Insulator A material, such as rubber or glass, that offers high resistance to electron flow.

Integrated Circuit Component containing diodes, transistors, resistors, capacitors, and other electronic components mounted on a single piece of material and capable to perform numerous functions.

Jumper Wire Wire used to temporarily bypass a circuit or components for electrical testing. A jumper wire consists of a length of wire with an alligator clip at each end.

Jump Start Procedure used when it becomes necessary to use a boost battery to start a vehicle with a discharged battery.

Kinetic Energy Energy in motion.

Laser Beam Alignment System A 2- or 4-wheel alignment system using wheel-mounted instruments to project a laser beam to measure toe, caster, and camber.

Lateral Runout The wobble or side-to-side movement of a rotating wheel.

Linkage System of rods and levers used to transmit motion or force.

Low-Maintenance Battery Conventionally vented, lead/acid battery, requiring normal periodic maintenance.

Maintenance-Free Battery Battery that does not require the addition of water during normal service life.

Maintenance Manual Publication containing routine maintenance procedures and intervals for vehicle components and systems.

National Automotive Technicians Education Foundation (NATEF) Foundation having a program of certifying secondary and post-secondary automotive and heavy-duty truck training programs.

National Institute for Automotive Service Excellence (ASE) Nonprofit organization that has an established certification program for automotive, heavy-duty truck, auto body repair, engine machine shop technicians, and parts specialists.

NIOSH National Institute for Occupation Safety and Health.

NLGI National Lubricating Grease Institute.

NHTSA National Highway Traffic Safety Administration.

NOP Nozzle Opening Pressure. Pressure an injector nozzle opens at in operation. Also known as VOP.

NOx Oxide of nitrogen compounds that may be formed during the combustion process. Collectively referred to as NOx. NOx compounds are responsible for the formation of photochemical smog.

OEM Original equipment manufacturer.

Off-road With reference to unpaved, rough, or ungraded terrain on which a vehicle will operate. Any terrain not considered part of the highway system falls into this category.

Ohm Unit of electrical resistance.

Ohm's Law Basic law of electricity stating that in any electrical circuit, current, resistance, and pressure work together in a mathematical relationship.

On-road With reference to paved or smooth-graded surface on which a vehicle will operate, part of the public highway system.

Open Circuit Electrical circuit whose path has been interrupted or broken either accidentally (a broken wire) or intentionally (a switch turned off).

Oscillation Movement in either fore/aft or side-to-side direction about a pivot point.

OSHA Occupational Safety and Health Administration.

Out-of-Round Eccentric.

Output Driver Electronic switch that the computer uses to control the output circuit. Output drivers are located in the output ECM.

Oval Condition that occurs when a tube is egg-shaped.

Overrunning Clutch Clutch mechanism that transmits power in one direction only.

Overspeed Governor Governor that shuts off fuel at a specific rpm.

Oxidation Inhibitor Additive used with lubricating oils to keep oil from oxidizing at high temperatures.

Parallel Circuit Electrical circuit that provides two or more paths for current flow.

Parallel Joint Type Type of driveshaft installation whereby all companion flanges and/or yokes in the complete driveline are parallel to each other with the working angles of the joints of a given shaft being equal and opposite.

Parking Brake Mechanically applied brake used to prevent a parked vehicle's movement.

Parts Requisition Form used to order new parts, on which the technician writes the part(s) needed along with the vehicle's VIN.

Payload Weight of the cargo carried by a truck, not including the weight of the body.

Pitting Surface irregularities resulting from corrosion.

PLN Pump-to-line-to-nozzle. Type of fuel injection system.

Polarity The state, either positive or negative, of charge differential.

Pole Number of input circuits made by an electrical switch.

Pounds per Square Inch (psi) Unit of English measure for pressure.

Power Measure of work being done factored with time.

Power Flow Flow of power from the input shaft through one or more sets of gears.

Power Train Engine to the wheels in a vehicle.

Pressure Force applied to a definite area measured in pounds per square inch (psi) English or kilopascals (kPa) metric.

Pressure Differential Difference in pressure between any two points of a system or a component.

Printed Circuit Board Electronic circuit board made of thin non-conductive material onto which conductive metal has been deposited. The metal is then etched by acid, leaving lines that form conductors for the circuits on the board. A printed circuit board can hold many complex circuits.

Programmable Read Only Memory (PROM) Electronic component that contains program information specific to vehicle model calibrations.

PROM Programmable Read Only Memory.

psi Pounds per square inch.

P-type Semiconductors Positively biased semiconductors.

RAM Random access memory. Main memory.

Ram Air Air forced into the engine housing or passenger compartment by the forward motion of the vehicle.

Random Access Memory (RAM) Memory used during computer operation to store temporary information. The microcomputer can write, read, and erase information from RAM. Electronically retained.

RCRA Resource Conservation and Recovery Act.

Reactivity Characteristic of a material that enables it to react violently with air, heat, water, or other materials.

Read Only Memory (ROM) Memory used in microcomputers to store information permanently.

Recall Bulletin Bulletin that pertains to special situations that involve service work or replacement of components in connection with a recall notice.

Reference Voltage Voltage supplied to a sensor by the computer, which acts as base-line voltage; modified by the sensor to act as an input signal. Usually 5 VDC.

Relay Electric switch that allows a small current to control a much larger one. It consists of a control circuit and a power circuit.

Reserve Capacity Rating Ability of a battery to sustain a minimum vehicle electrical load in the event of a charging system failure.

Resistance Opposition to current flow in an electrical circuit.

Resource Conservation and Recovery Act (RCRA) Law that states that after using hazardous material, it must be properly stored until an approved hazardous waste hauler arrives to take it to a disposal site.

Revolutions per Minute (rpm) Number of complete turns a shaft turns in one minute.

Right to Know Law Law passed by the federal government and administered by the Occupational Safety and Health Administration (OSHA) that requires any company that uses or produces hazardous chemicals or substances to inform its employees, customers, and vendors of any potential hazards that may exist in the workplace as a result of using the products.

Ring Gear Gear around the edge of a flywheel.

ROM Read only memory.

Rotary Oil Flow Condition caused by the centrifugal force applied to the fluid as the converter rotates around its axis.

Rotation Term used to describe a gear, shaft, or other device when it is turning.

Rotor Rotating part of the alternator that provides the magnetic fields necessary to create a current flow. The rotating member of an assembly.

rpm Revolutions per minute.

Runout Deviation or wobble of a shaft or wheel as it rotates. Measured with a dial indicator.

Semiconductor Solid state material used in diodes and transistors.

Sensing Voltage Voltage that allows the regulator to sense and monitor the battery voltage level.

Sensor Electronic device used to monitor conditions for computer control requirements. An input circuit device.

Series Circuit Circuit connected to a voltage source with only one path for electron flow.

Series/Parallel Circuit Circuit designed so both series and parallel conditions exist within the same circuit.

Service Bulletin Publication that provides the latest service tips, field repairs, product improvements, and related information of benefit to service personnel.

Service Manual A manual, published by the manufacturer, that contains service and repair information for all vehicle systems and components.

Short Circuit Undesirable connection between two worn or damaged wires. The short occurs when the insulation is worn between two adjacent wires and the metal in each wire contacts the other, or when wires are damaged or pinched.

Single-Axle Suspension Suspension with one axle.

Single-Reduction Axle Any axle assembly that employs only one gear reduction through its differential carrier assembly.

Solenoid Electromagnet used to conduct electrical energy in mechanical movement.

Solid Wires Single-strand conductor.

Solvent Substance that dissolves other substances.

Spade Fuse Term used for blade fuse.

Spalling Surface fatigue that occurs when chips, scales, or flakes of metal break off.

Specialty Service Shop A shop that specializes in areas such as engine rebuilding, transmission/axle overhauling, brake, air conditioning/heating repairs, and electrical/electronic work.

Specific Gravity Ratio of a liquid's mass to that of an equal volume of distilled water.

Spontaneous Combustion Reaction in which a combustible material self-ignites.

Stall Test Test performed when there is a malfunction in the vehicle's power package (engine and transmission), to determine which of the components is at fault.

Starter Circuit Circuit that carries the high current flow and supplies power for engine cranking.

Starter Motor Device that converts electrical energy from the battery into mechanical energy for cranking.

Starter Safety Switch Switch that prevents vehicles with automatic transmissions from being started in gear.

Static Balance Balance at rest, or still balance.

Stepped Resistor Resistor designed to have two or more fixed values, available by connecting wires to one of several taps.

Storage Battery Battery to provide a source of direct current electricity for both the electrical and electronic systems.

Stranded Wire Wire that is made up of a number of small solid wires, generally twisted together, to form a single conductor.

Sulfation Condition that occurs when sulfate is allowed to remain on the battery plates for a long time, causing two problems: (1) it lowers the specific gravity levels, increasing the danger of freezing at low temperatures; and (2) in cold weather a sulfated battery may not have the reserve power needed to crank the engine.

Swage To reduce or taper.

Switch Device used to control on/off and direct the flow of current in a circuit. A switch can be under the control of the driver or can be self-operating through a condition of the circuit, the vehicle, or the environment.

Tachometer Instrument that indicates shaft rotating speeds.

Throw (1) Offset of a crankshaft. (2) Number of output circuits of a switch.

Time Guide Used for computing compensation payable by the truck manufacturer for repairs or service work to vehicles under warranty.

Timing The phasing of events to produce action such as ignition.

Torque Twisting force.

Torque Converter Device, similar to a fluid coupling, that transfers engine torque to the transmission input shaft and can multiply engine torque.

Toxicity Statement of how poisonous a substance is.

Tractor Motor vehicle that has a fifth wheel and is used for pulling a semi trailer.

Transistor Electronic device produced by joining three sections of semiconductor materials. Used as a switching device.

Tree Diagnosis Chart Chart used to provide a logical sequence for what should be inspected or tested when troubleshooting a repair problem.

Vacuum Pressure values below atmospheric pressure.

Vehicle Retarder Engine or driveline brake.

VGT Variable geometry turbocharger. A turbocharger with moveable vanes in the turbine section of the turbocharger, used to regulate exhaust gas flow. Can control boost pressure and turbocharger speed. Also referred to as VNT (variable nozzle turbocharger).

VIN Vehicle Identification Number.

Viscosity Resistance to flow or fluid sheer.

Volt Unit of electromotive force.

Voltage-Generating Sensors Devices that produce their own voltage signal.

Voltage Limiter Device that provides protection by limiting voltage to the instrument panel gauges to approximately 5 volts.

Voltage Regulator Device that controls the current produced by the alternator and thus the voltage level in the charging circuit.

VOP Valve Opening Pressure. Caterpillar term for NOP.

Watt Measure of electrical power.

Watt's Law Law of electricity used to calculate the power consumed in electrical circuit, expressed in watts. It states that power equals voltage multiplied by current.

Windings (1) Three bundles of wires in the stator. (2) Coil of wire in a relay or similar device.

Work (1) Forcing a current through a resistance. (2) Product of a force.

Yield Strength The highest stress a material can stand without permanent deformation or damage, expressed in pounds per square inch (psi).

Notes

Notes

Notes

Notes

Notes

Notes

Notes